Digital Hermeneutics

This is the first monograph to develop a hermeneutic approach to the digital—as both a technological milieu and a cultural phenomenon. While philosophical in its orientation, the book covers a wide body of literature across science and technology studies, media studies, digital humanities, digital sociology, cognitive science, and the study of artificial intelligence.

In the first part of the book, the author formulates an epistemological thesis according to which the "virtual never ended." Although the frontiers between the real and the virtual are certainly more porous today, they still exist and endure. In the book's second part, the author offers an ontological reflection on emerging digital technologies as "imaginative machines." He introduces the concept of emagination, arguing that human schematizations are always externalized into technologies, and that human imagination has its analog in the digital dynamics of articulation between databases and algorithms. The author takes an ethical and political stance in the concluding chapter. He resorts to the notion of "digital *habitus*" for claiming that within the digital we are repeatedly being reconducted to an oversimplified image and understanding of ourselves.

Digital Hermeneutics will be of interest to scholars across a wide range of disciplines, including those working on philosophy of technology, hermeneutics, science and technology studies, media studies, and the digital humanities.

Alberto Romele is Associate Professor of Philosophy of Technology at Lille Catholic University, France. He is the co-editor of *Towards a Philosophy of Digital Media* (2018).

Routledge Studies in Contemporary Philosophy

For more information about this series, please visit: www.routledge.com/
Routledge-Studies-in-Contemporary-Philosophy/book-series/SE0720

Digital Hermeneutics
Philosophical Investigations in
New Media and Technologies

Alberto Romele

NEW YORK AND LONDON

First published 2020
by Routledge
52 Vanderbilt Avenue, New York, NY 10017

and by Routledge
2 Park Square, Milton Park, Abingdon, Oxon OX14 4RN

Routledge is an imprint of the Taylor & Francis Group, an informa business

First issued in paperback 2021

Library of Congress Cataloging-in-Publication Data
A catalog record for this title has been requested

ISBN: 978-0-367-35366-7 (hbk)
ISBN: 978-1-03-208817-4 (pbk)
ISBN: 978-0-429-33189-3 (ebk)

Typeset in Sabon
by Apex CoVantage, LLC

Contents

Acknowledgments

Chapter 3 is a reworked version of Romele, A. 2018. "Imaginative Machines." *Techné: Research in Philosophy and Technology* 22(1): 98–125. Chapter 4 is a reworked version of Romele, A. 2018. "From Registration to Emagination." In A. Romele, and E. Terrone (eds.). *Towards a Philosophy of Digital Media.* London: Palgrave Macmillan, 257–273. The introduction and the conclusion of part 2 are a reworked version of Romele, A. 2019. "Towards a Posthuman Hermeneutics." *Journal of Posthuman Studies* 3(1): 45–59.

I started thinking about this book in Italy. I wrote it largely in Portugal and I finished it in France. I decided to write it in English, which is my "no man's land." On this long journey, I met many people who taught, helped, and supported me in different ways. I would like to especially thank Mario Lombardo, Jos De Mul, Marta Severo, Christian Berner, Ernst Wolff, Marcel Hénaff, Stefano Gualeni, Jean-Claude Gens, Bruno Bachimont, Johann Michel, Rossana de Angelis, Franck Cormerais, Paolo Furia, Camilla Emmenegger, Francesco Gallino, Claudio Paravati, Daniele Gorgone, Enrico Terrone, Dario Rodighiero, François-David Sebbah, Luca Possati, Wessel Reijers, Mark Coeckelbergh, and Maurizio Ferraris.

I would like to express my gratitude to my former postdoctoral supervisor Paulo Tunhas and my old colleagues at the Institute of Philosophy of the University of Porto: José Meirinhos, Eleonora Lombardo, João Rebalde, José Higuera, Celia López, Paula Silva, Isabel Marques, Sofia Miguens, and Mattia Riccardi. My heartfelt thanks also to the Fundaçao para a Ciência e a Tecnologia that financed my postdoctoral researches (grant number SFRH/BPD/93297/2013). At the ETHICS (EA-7446) lab of the Catholic University of Lille, I have appreciated the tremendous support given to me by my new colleagues, in particular, Nicolas Vaillant, Malik Bozzo-Rey, Alain Loute, and the members of the research chair "Ethics, technology, and transhumanisms": David Doat, Benjamin Bourcier, Stanislas Deprez, Paul Jorion, Jacques Printz, Vincent Calais, Marie-des-Neiges Ruffo, Gabriel Dorthe, Fernand Doridot, and Franck Damour. Jyotsna Massey and Sara Testa helped me a lot with the English grammar.

I am also grateful to all my friends, especially Alessandro, Mattia, Ilya, Davide, Enzo, Marco, Laura, Alice, Elisa, Sebastian, Eleonora, and Sara. I owe everything to my family of origin: my parents Nadia and Luigi, and my sister, Elena, who gave me two fantastic nieces, Lavinia and Beatrice, a newborn nephew, Tommaso, and even an acceptable brother-in-law, Michele. Finally, it would not be enough to *thank* the family of which I am now part of, Marta, Matteo, Carlo, and Clara. I have no words to describe the joy they give me every day.

Overture

The Idealism of Matter

I want to introduce this book with a crude confession. When I started to conduct research on hermeneutics and digital technologies, my intentions were mostly destructive. In a previous book of mine (Romele 2013a), a small part of which has been published in English (Romele 2014), I had already worked on a deconstruction of philosophical hermeneutics 'from above.' That work was built on Gadamer's rehabilitation of the Augustinian notion of inner word (*verbum in corde*, literally the "verb in the heart") in the third part of *Truth and Method*. On the one hand, like most of the interpreters—such as Grondin (1993)—I agreed that the Augustinian inner word was the paradigm of ontological hermeneutics' claim to universality. On the other hand, however, I disagreed with them because my thesis was that the Augustinian inner word does not show the potentialities, but rather the ineffectiveness, of ontological hermeneutics. The *verbum in corde* does not suggest that there is a "dialogue" behind all utterances. It indicates that behind all dialogue there is just a monologue of what Christians call God and Heidegger preferred to refer to as Being.

I have, for instance, demonstrated that over the years Augustine devaluates more and more human 'horizontal' speech and written word, along with their communicational potential, in favor of a 'vertical' revelation of truth given by divine grace. I have also demonstrated that despite his declared plea for the dialogue, Gadamer remains, in the end, faithful to the evenemential (*événementiel*) and anti-communicational model developed by Augustine and mediated by both protestant theology and the late Heidegger. For Gadamer, interpretation does not have any intrinsic value. I was certainly caricaturing some aspects of this tradition, but my impression was that for ontological hermeneutists, interpretation is no more than a long training for achieving the correct disposition to passively receive the sole truth of the Event. My reaction to such a perspective did not consist in putting the Being "under erasure (*sous rature*)," but in merely eliminating it, in favor of a minor and pragmatic hermeneutics for which correct interpretations exist but always depend on the context and are for this reason, in principle, renegotiable.

My initial investigations into the digital aimed at completing this research by deconstructing, this time, ontological hermeneutics 'from below.' In some sense, I wanted to pull the rug out from under this tradition; or even better, I intended to show that there was no rug, no ground, and no other kind of matter under the feet of hermeneutists such as Gadamer, Heidegger, and Ricoeur. One could say that within ontological hermeneutics there is too much of presence above, and too much of absence below. My hypothesis was that, despite their declared interest for language in all its forms, these hermeneutists were victims of what might be called an 'idealism of matter.' Bruno Latour (2007) refers to "idealist materialism" as the tendency in the past to consider things exclusively in their primary qualities. Such a form of materialism, Latour says (139), "ignores entirely the difficulty of producing drawings and the whole network of engineering practices necessary to identify the features, follow the lines, and assemble the whole institution necessary for any mechanism to exist." He takes the example of Heidegger's philosophy of technology and wonders how it is possible to take his approach seriously. How can one believe that different technologies such as an atomic bomb, a dam, a lie detector, and a staple are mere varieties of the same "enframing"? How can one reduce a hammer or a shoe to an assembly of just four elements (the Heideggerian "fourfold")?

In the case of ontological hermeneutics dealing with language, the idealism of matter is twofold. Firstly, among the several mediators between the human and the world, hermeneutics has considered language alone. This reflects the tendency of all philosophies and other human and social sciences of the "linguistic turn." According to Latour (1993, 63), the greatness of the philosophies and theories of language of the twentieth century is that they gave mediators a proper dignity, in the sense that they were no longer considered as pure vehicles conveying meaning from the speaker to the world, or vice versa. However, their weakness is that they elected language as the sole mediator, or at least as the paradigm of all possible mediations.[1] Secondly, the language philosophical hermeneutics deals with is, so to say, spirit without matter. Even in its most materialized forms, language has been treated as if it were a bodiless soul. The mediators have been idealized because they have been understood through the prism of an idealized language.

Consider Heidegger's philosophy of technology. I am thinking here firstly of his reflections on the "enframing" and the "fourfold." Determinism, pessimism, and 'monolithism' (in the sense of a discourse on Technology with a capital "T")—these are the three aspects that characterize the considerations one can find, for instance, in "The Question Concerning Technology." According to Carl Mitcham (1994, 50), it is significant that of the three works Heidegger titled "The Question of . . .," one is about Technology, one is about Being, and another one about the Thing. This suggests, according to him, a need to examine the "question concerning Technology," especially in relation to the "question of Being"—and maybe even the "question of the Thing": "Modern technology not only covers over or obscures the thinghood in things, it also covers over or obscures the Being of beings, and ultimately itself. Technology cannot be understood in terms of technology" (53). This means that from the Heideggerian perspective, Technology is the adversary of Being—or the Being that gives itself in its concealment.[2] For this reason, one can argue that Technology has the same effects on the understanding of technologies that Being has on the understanding of the plurality of beings. As I mentioned earlier, in the end, in ontological hermeneutics interpretation has no intrinsic value, but is rather to be understood as a long training for acquiring the correct disposition for passively receiving the Truth-Event. Similarly, the few descriptions of modern technologies offered by Heidegger have the sole function of referring to the unique logic of "enframing." For Heidegger, contrary to the ancient *technè*, Technology obscures reality, because it approaches it as raw material at disposal to be used and exploited. However, one could say that the Heideggerian notion of Technology paradoxically ends up obscuring the plural reality of technologies. The notion of Being conceals the interpretation and understanding of a multitude of beings, as the notion of Technology conceals the interpretation and understanding of a multitude of technologies and technological processes. The ontological attitude in hermeneutics has the effect of anesthetizing, so to speak, the interest for the multiple and contextual ontic interpretations. Similarly, the ontological monolithism in "humanities philosophy of technology" anesthetizes the ontic interest for concrete technological ensembles.

But, in this case, the solution is rather simple: one has to eliminate the problematic notion (once it has been demonstrated that such notion is questionable), and the "material materialism" and the possibility of a "thick description of things" will start to appear. Was it necessary to "raise anew *the question of being*" as Heidegger has imposed on to his readers at the very beginning of *Being and Time*? Do we need to raise any question about Technology? Considering the enormous simplifications and misunderstanding this attitude has provoked, I doubt it. And it is rather surprising how easy it can be to say that "the question of Being" and "the question concerning Technology" were just the wrong questions to ask.

Heidegger's "turn toward things" is certainly more interesting in this respect. According to Peter-Paul Verbeek (2005, 76), the most promising point of departure for a hermeneutic philosophy of technology is found in Heidegger's early work, and more specifically in his analysis of tools in *Being and Time*. One can certainly have a good and prosperous—even philosophical—life while ignoring invisible entities like Being and Technology. Yet for several obvious reasons, it is harder to ignore the plethora of beings and technologies that we deal with in our everyday life. Verbeek insists on the fact that in Heidegger's analysis of "being-in-the-world" tools play an important role because they make it possible for relations between humans and the world to come about. He specifically refers to the concept of *Zuhandenheit*, "handiness" or "readiness-to-hand." Generally, humans do not focus on the tool they are using, but on the work or task in which they are engaged. Tools hide in the relation between human beings and the world, but they also contribute fundamentally to shape the world as it appears to us: "Things, in short, disclose the world. When somebody uses a tool or piece of equipment, a referential structure comes about in which the object produced, the material out of which it is made, the future user, the environment in which it has a place are related to each other" (79). Yet, such a complex structure explicitly appears just when a handy tool breaks down. In this case, the tool itself demands attention, and the formerly transparent chain of references becomes problematic. The tool changes over from being "ready-to-hand" to be *Vorhanden*, that is, "present-at-hand" or "objectively present."

Every researcher in philosophy of technology is familiar with this story and heard about the Heideggerian example of the hammer. My thesis is that in these passages Heidegger is not interested in technologies per se. Neither is he interested in the primacy of the practical attitude, as argued, for instance, by Hubert Dreyfus. The tool and the process that goes from its daily use to its sudden break has the sole function to bring the *Dasein* toward an authentic form of self-understanding. It is indeed true, then, that at the beginning of § 16 of *Being and Time* Heidegger says that "when we discover its unusability, the thing becomes conspicuous" (Heidegger 1996, 68). It is also true that he affirms, in § 18, that "the total relevance which, for example, constitutes the things at hand in a workshop in their handiness is 'earlier' than any single useful thing, as is the farmstead with all its utensils and neighboring lands" (78). However, immediately after he recalls that "the for-the-sake-of-which always concerns the being of *Da-sein* which is essentially concerned *about* this being itself in its being" (ibid.). In other words, the material understanding of the constitutive role of tools in the practical mediation between humans and the world is always submitted to an ideal series of considerations concerning the possibility for the *Dasein* to authentically understand oneself.

Two more observations are necessary. The first one is that the paradigm of the "handiness" is still language and a specific interpretation of it. In order to understand tools, Heidegger uses the linguistic notion of "reference (*Verweisung*)." As the sign in Saussurian structural linguistic, the tool has for him just a differential value: "Strictly speaking, he says in § 15 (64), there 'is' no such thing as *a* useful thing. There always belongs to the being of a useful thing a totality of useful things in which this useful thing can be what it is." Moreover, when it comes to "grasping the phenomenon of *reference* more precisely (*schärfer*)," as Heidegger says to be the case in § 17, he completely puts aside tools, and starts to consider signs, which he defines as "useful things whose specific character as useful things consists in *indicating*" (72). The second consideration is that the kind of language Heidegger is truly interested in is a sort of inner monologue of the *Dasein* with oneself. So just as tools are reduced to their most abstract form (that is, signs), in the same way, signs are reduced to their most abstract version, that is, the inner word of Conscience. Two elements, at least, characterize the call of Conscience—first, that "*Conscience speaks solely and constantly in the mode of silence*" (252); second, the fact that in Conscience, *Dasein* calls itself, in a sort of self-awakening process toward his or her most authentic authenticity. In sum, Heidegger's early "turn towards things" is always-already turned away from things, since it understands things in the light of language, and language in the light of its most idealized form, which is the call of Conscience. I am not arguing that one cannot extrapolate a philosophy of technology *from* Heidegger by isolating some passages of *Being and Time* and other works. I am instead arguing against the possibility of developing any kind of empirical philosophy of technology *within* a Heideggerian framework.

* * *

I want now take this argument a step further. For the Heidegger of *Being and Time*, language is the tool par excellence, and one's monologue with oneself is the paradigm for discriminating between authentic language and "idle talk." In § 34, discourse (*Rede*) is defined as "the 'significant' articulation of the intelligibility of being-in-the-world" (151). And yet, in the same paragraph, the centrality of listening and keeping silent is also stressed: "In talking with one another the person who is silent can 'let something be understood', that is, he can develop an understanding more authentically than the person who never runs out of words" (154). In other words, the authentic dialogue, which is the model of our dealing with the world, finds its truth in the passive openness to Being. In this movement from the ontic to the ontological dimension, language loses most of its materiality, and it is precisely for this reason that one cannot resort to Heideggerian philosophy to develop a material understanding of technologies. Don Ihde (2010, 117) quite recently came to a similar conclusion: "Heidegger does remain a giant mighty dead twentieth-century philosopher—but *not* with respect to the philosophy of technology!" In particular, he quotes Graham Harman, according to whom for Heidegger "[a] mass-produced umbrella is no different from a cinder block or an aircraft carrier. [. . .] [T]he problem with his analyses is not their pessimism, but their monotony" (in Ihde 2010, 114). Such monotony is already present in his earlier reflections on tools, however differently. While in his works, which are explicitly devoted to Technology, he immediately considers tools as an undifferentiated whole, in his earlier investigations he comes to the same conclusion by degrees, making all tools merge into language, and making all languages come together in a form of monologue.

Still, I have never believed that deconstructing Heidegger is of exceptional merit. The limits of his approach are evident to the majority of his interpreters. For me, a more interesting challenge was to work on the ontological hermeneutics of Paul Ricoeur. Among the hermeneutists, Ricoeur is indeed the one who has most insisted on the externalizations and materializations of language. Ricoeur has focused his attention on written and fixed forms of language: symbols, signs, narratives, et cetera. In *From Text to Action*, he stresses, for instance, the ontological priority of writing over speech: "The psychological and sociological priority of speech over writing is not in question. It may be asked, however, whether the late appearance of writing has not provoked a radical change in our relation to the very statements of our discourse" (Ricoeur 1991, 106). Ricoeur shares here the thesis one can find for example in Derrida, but also in authors such as Walter G. Ong and Jack Goody. It is precisely for this reason that his hermeneutics is potentially interesting for the philosophy of technology. There is, in fact, a certain interest in Ricoeur as a philosopher of technology. Ernst Wolff (2013) has distinguished among three main research directions in this field. The first one is that of David Kaplan (2006), who investigated what Ricoeur effectively said about technology and proposed then to develop a "Ricoeurian critical theory."[3] The second one is David Lewin's (2012), who tried to show how Ricoeur's thought could be relevant in a debate à la Heidegger on Being and Technology. The third one is the work of Wolff himself, who has been using Ricoeur's hermeneutics and philosophical anthropology to develop an interdisciplinary approach on the technical dimension of human action. In addition to these approaches, I would add the contributions of authors like Bruno Gransche (2017), Mark Coeckelbergh and Wessel Reijers (2016), and myself, which consist in resorting to Ricoeur's hermeneutics in the specific context of digital technologies.

However, before using Ricoeur's philosophy positively, as it will be the case in some place in this book, I would like to continue to present my initial work of deconstruction. On the one hand, Ricoeur externalizes and materializes language and interpretational processes, in general, more than all other ontological hermeneutists. On the other hand, however, his understanding of the materialized language remains paradoxically idealistic. Ricoeur is particularly interested in the semiotic and semantic rules that underlie the functioning of signs, symbols, linguistic metaphors, and narratives. Yet, he is not interested into the materialities of these means for transmitting meaning. It is precisely for this reason that digital technologies have started to become interesting for me, insofar as they invite one to consider materialities other than those of traditional media such as printed books.

In Romele (2013b) I have used social networking sites to reconsider some assumptions of the Ricoeurian concept of narrative identity, which is characterized by both mono-linearity and mono-mediality. Such a notion indicates that narratives, that is, the stories we read, tell, and listen to about others and ourselves, have heuristic effects on our lives. It is also a concept at the frontier between description and prescription according to which a fundamental part of our identities is or should be constituted narratively. Now, this notion is mono-linear because it is based on an Aristotelian (the one of the *Poetics*) and biblical model, according to which all narratives must be composed of an "emplotment" (*mise en intrigue* in French), which brings the heterogeneous elements of a situation in an imaginative order, enclosed between a beginning and an end. For him, even contemporary (during his time) experimental novels do not transgress such a model—firstly, because in order to transgress it they also have to bear a certain trace of it; secondly, because in cases like Robbe-Grillet's *nouveau roman* the emplotment is not abolished but is simply postponed at the moment of reading. The mono-mediality of this concept depends on same reason, i.e. that it is built on the model of the printed book.[4] According to Jos De Mul (2010), while the reading of the printed text tends to be univocal, the digital text is potentially different from every single reading. The Web is for him a database of an indefinite number of potential stories. Moreover, in the digital, the written form is just one of a series of media and possibilities of expression. In another article (De Mul 2015), the Dutch philosopher compares the narrative identity with the ludic identity, a notion built on the paradigm of computer games. Whereas the former looks back to the past, a story to be told that has already happened, the latter is geared toward the future, as a game to be played without the player getting bored must not have been played yet. While stories are enclosed in an emplotment already decided by the author, computer games allow the player to move freely within a framework of few given rules. In some sense, computer games are the concrete realization of the speculations of postmodernity and the clumsy attempts of some literary avant-gardes.

Digital technologies represent alternative materialities that I have used to dismantle the idealism of matter—or the idealized matter—which represents the basis of many Ricoeurian theories about interpretation. On the one hand, Ricoeur is certainly conscious of the importance of externalization and materialization. This is the sense, for instance, of his "long route" opposed to the Heideggerian "short route." Interpretation is not immediately oriented to the self, but the self is reached through a deviation among the externalizations and the methods, techniques, and technologies. Such is more generally our experience of the world, which is never immediate or direct as in the Husserlian idealist phenomenology, but always (technologically) mediated. On the other hand, however, he indiscriminately resorts to a specific kind of externalizations for understanding all the others. The same holds true for what concerns the methods because Ricoeur tends to understand all methodologies in human and social sciences in the light of linguistic structuralism. But such attitude has ruinous effects on the possible understanding of the self. For this reason, authors like Bouchardon (2014, 146–147) and De Mul have insisted on that digital multi-linearity and multi-mediality might be better paradigms (certainly less frustrating) for approaching our postmodern identities which have the form of patchworks, bricolages or montages of different and not always coherent with each other's elements.[5] There are several reasons to disagree with this 'joyful' understanding of the digital. I will account for many of them in this book. However, this is not the point here. The point is that ontological hermeneutics, a discipline that gave much importance to language and mediation in general, demonstrates to have not thought about language and mediation with enough depth. Its considerations remain superficial, certainly not for lack of abstraction, but rather for an excess of it. They are superficial because ontological hermeneutics has refused to dig the earth, to dirty its own hands with the matter of which language is always made of. Ricoeur is for me an interesting example because he represents at the same time the glory and the failure of this discipline: even in the most externalized and materialized expressions of language, he has been unable to see the fundamental role matter and technologies have in the variations of the meaning transmission.

In another article (Romele 2018), I have analyzed the idealized way in which Ricoeur presents the archive in *Memory, History, Forgetting*. In this case, too, Ricoeur takes a material and technical object, namely the historical archive constituted by written documents, and universalizes it. In this way, however, he cannot understand the potential of other forms of archiving for understanding the historical reality such as those allowed by audiovisual and digital technologies. For the French philosopher, the archive is an essential moment in the dialectic (the hermeneutic circularity) between the veracity of memory and the truth of history, in which "things said tip from the oral field to that of writing, which history will not henceforth abandon" (Ricoeur 2004a, 146). The archive, he says, "is the moment of entry into the writing of the historiographical operation. Testimony is by origin oral. It is listened to and heard. The archive is written. It is read and consulted. In archives, the professional historian is a reader" (166). Maybe, by using the words "orality" and "writing," Ricoeur wants to say something more than what is usually meant with these two terms. Yet, it is precisely this approximation, this flattening the archive on writing that impedes him to see that the historical truth tends to change in form and contents with the changing of the techniques and technologies at disposal. If technologies essentially determine the way the veracity of memory is taken and transported into the truth of history, then no hermeneutics of the historical knowledge can do without reflecting on the technologies available and their evolution. The nature of some historical and cultural phenomena imposes other kinds of archiving than the mere writing down. This is, for instance, the case of the intangible cultural heritage, that is, according to the UNESCO definition, "traditions or living expressions inherited from our ancestors and passed on to our descendants, such as oral traditions, performing arts, social practices, rituals, festive events, knowledge and practices concerning nature and the universe or the knowledge and skills to produce traditional crafts."[6]

It would be undoubtedly naïve to think that the digital alone is enough to resolve the issue of safeguarding such a complex object, which seems unstable by nature. Marta Severo (2018) has recently highlighted the paradoxical nature of any attempt of "safeguarding without a record." However, again, this is not the point here. The point is rather that the example of digital technologies supports me in arguing against the idealized matter that characterizes most of Ricoeur's reflections on hermeneutics. In other words, digital technologies soften the boundaries between the lifeworld and its hermeneutical understanding. In another text, the French philosopher suggests that lifeworld and the idealities of science are irreducible to each other: "This distinction [. . .] suggests the idea of an irreducible dialectic [. . .] between the real world, as soil, and the idea of science, as the principle of all validation. [. . .] We can 'derive' the idealities, in the sense that they are referring to the real world. But we cannot derive their quest for truth" (Ricoeur 2004b, 377).[7] Ricoeur is discussing here the Husserlian "backward questioning," but one can say that what the later Husserl has tried to do in the field of hard sciences, Ricoeur suggested doing in the context of human and social sciences (Michel 2015, 24). On the one hand, I entirely agree with such irreducibility, because there is and there will be no technique or technology that will make us capable of overcoming such a gap. This is in part due to technological limitations, but also due to our cognitive limitations. These points will be extensively discussed in the rest of the book. Yet, on the other hand, I have the impression that Ricoeur and most of the classic hermeneutists tend to treat the frontiers between the lifeworld and its understanding as if they were stable, but instead they sensibly vary when the sociotechnical conditions change. The digital can be precisely used to show the relevance of these flexibilities.

* * *

However, one has to admit that the idealism of matter is not an intrinsic limit of hermeneutics. For instance, Don Ihde (1998) has worked hard for "expanding hermeneutics"—firstly, by renouncing the primacy of language: against the "literalization" of the world, i.e. the fact that in hermeneutics as well as in other philosophies of the linguistic turn everything turns out to be "textlike," he insisted on other aspects like perception and embodiment. Secondly, he has analyzed in depth the ways technologies constitutively mediate between the world and our understanding of it. In other words, Ihde has the merit to have deployed those technological flexibilities within a hermeneutic framework that in traditional hermeneutics were still transparent or invisible. Interestingly enough, one might distinguish between a 'special' and a 'general' hermeneutics in Ihde's thought. On the one hand, he calls "hermeneutic" a specific kind of technologically mediated human-world relations, in which the technology offers a representation of the world that must be 'interpreted' to access the world. On the other hand, he implicitly recognizes that all technologies are hermeneutic 'by nature' when he discusses their "magnification-reduction structure," namely that they are selective—by stressing some aspects of the world and neglecting some others—and do not simply replicate non-technological situations. One may argue that in the philosophy of technology it is always a matter of interpretation, in the sense of distributing the right amount of interpretational agency among humans and machines, an amount that can considerably vary from close-to-zero importance of the machine to an inversion of roles.

It is precisely at this point that I have discovered, in the course of my research, the heuristic potential of hermeneutics for understanding the digital. Digital technologies are indeed hermeneutics both in a 'special' and in a 'general' sense; in a special sense, they are hermeneutics because they structurally have and offer representations of the world that must be interpreted to access the world, and in a general sense, because in this field one has to be able to discriminate between the interpretative agencies of humans and non-humans. While in my initial investigations I resorted to the digital in order to deconstruct hermeneutics, this book mainly focuses on the use of hermeneutics for understanding some aspects of the digital. If "hermeneutics *of the* digital" were the title of this book, one would say that whereas I have entered this research field through the subjective use of the genitive, I have ended up developing an investigation focusing on the objective use of the genitive.

Not only digital technologies are structurally hermeneutic, but I would also say that there is no hermeneutics which is more effective than digital hermeneutics because in digital hermeneutics writing, signs, and symbols do not limit themselves to represent the world. In representing it, they also modify several aspects of it. Digital symbolism has an operational character that lacks classic linguistic and other forms of symbolism. While in magic or religion the effectiveness of symbols remains a form of unrealizable desire or belief—indeed with significant cognitive and social consequences—digital symbolism concretely performs what it promises. It is important to notice how such effectiveness of the digital is not limited to the digital world. The interest philosophers and theorists in general have shown in recent years for technologies such as the Internet of Things, 3-D printing, drones, et cetera is precisely due to the fact that these technologies actualize the digital. More radically, these technologies are self-actualizations of the digital in the world, and actualizations of the world through the digital. This, of course, does not mean that algorithmic trading or other digital processes that remain digital from the beginning to the end and are never actualized are less relevant to our lives. It just means that in some cases the effectiveness of the digital becomes more visible and its consequences are immediately tangible.

However, there is also a fundamental difference between hermeneutics and the digital. Since the digital realizes what it says—a virtue which is rare among human beings—there is a sociocultural tendency to believe that the digital can realize everything. There is also a tendency to reduce the world to its digital representations, that is, to neglect the distances and the discrepancies between the world and its digitally mediated understandings. Such a distance, or distanciation, plays, instead, a central role in philosophical hermeneutics, at least in the hermeneutics of Paul Ricoeur. While Gadamer has privileged the immediate "fusion of horizons," Ricoeur used an externalized and materialized form of language, namely the text, to reintroduce a positive or even productive notion of distanciation. The French philosopher particularly insists on the autonomy of the text before its author. He also insists on the separation between the meaning and the reference, that is, between the representation and its truth value, its claim to reach reality. For him, the abolition of the reference in literature paves the way to a second and more fundamental reference, which has to do with the reconfigurative effects of the texts we read on our existences.

Distanciation plays a fundamental role in the digital as well. Indeed, the digital is effective into the world because it 'translates' or 'transcodes' it. Yet, the digital is surrounded by an illusion of plenitude. For Ricoeur, the heuristic value of literature depends precisely on its distanciation from life, in the sense that novels offer *alternative* models and possibilities to one's life. Incidentally, I would be less optimistic about what concerns literature. Firstly, because it can be easy to confuse literature for reality; it is well known, for example, that novels such as *Madame Bovary* and *Anna Karenina* provoked an increase in suicides in the years following their publications. Secondly, because literature can have effects that are different from the reconfiguration for the better; for instance, it can induce forms of coercion and standardization. Let us think back to some religious books (or at least to some ways to understand and use them) or to the fact that literature is no less subject to the effects of the cultural industry attaining other arts. However, I am not interested in this context to criticize Ricoeur's "semantic optimism" (Loute 2017). I want to rather insist that the digital, both as a technological and a cultural phenomenon, tends to flatten the world on *what it already is*. Thus, the heuristic scope of a digital hermeneutics consists in practicing a form of "backward questioning" of the digital, that is, in insisting on the hermeneutic distance, but also circularity, between the lifeworld and its digital understandings. I would like to stress that such a critical attitude is different from the external and extrinsic critique that several humanists have been making of the digital. My approach is rather pharmacological (Stiegler 2014), in the sense that for me it is a matter of hermeneutically articulating limits and possibilities of the digital, its risks and opportunities—*pharmakon* is a Greek word meaning both "poison" and "remedy."

The first part of the book is a sort of pharmacology of the digital. I will start by resorting to the provocative motto, an homage to Philip K. Dick, according to whom "the virtual never ended." Such a statement is provocative insofar as it rehabilitates the old-fashioned notions of virtual and virtuality, as well as the distinction between the virtual and the real that dominated the discourses on the digital during the 1980s and the 1990s. My thesis is that although the frontiers between these two dimensions are certainly a lot more porous today, they still exist and resist. For this reason, in chapter 1 I will take a position against those who believe that 'the virtual invaded the real.' In particular, I will deal with Luciano Floridi's semantic theory of information (STI) and some of its ethical and ontological consequences. My idea is that such a theory brings to what might be said to be 'a night in which all cows are informational.' As opposed to it, I will propose a hermeneutic approach to information, which is a sort of a third way between those information theories that exclude in principle any semantic and pragmatic implication and those that instead include them from the beginning. In the chapter 2, I will criticize those for whom 'the real has invaded the virtual.' More specifically, I am going to discuss some recent results in the field of digital sociology, a discipline according to which digital traces are, if opportunely analyzed, faithful representations of the social reality. A long paragraph will be devoted to Bruno Latour and his few, but rather influential, considerations on the digital. My hypothesis here will be that his perspective on the digital contradicts his philosophical and sociological insights. Such a contradiction reveals the partial inconsistency of his version of the actor-network theory. In the conclusion, I will more specifically discuss the existing literature on digital hermeneutics and deliver some remarks on the concept of digital trace. In the phenomenological tradition, the notion of trace refers to the "absence of a presence," i.e. to an essential gap between the meaning and its reference. For this reason, I am going to suggest to use this concept in the field of digital studies as an alternative to the well-formed notion of information.

While the first part of the book is devoted to the epistemological and methodological issues in digital hermeneutics, in the second part, I will undertake a sort of ontological turn. This expression does not refer here to a radical orientation toward Being, but more modestly to an interest for the emerging capacities of some digital machines. Latour (1994) has described a spectrum of technological mediations that goes from "translation" to "delegation." Translation happens whenever in the use of technology a new goal is created that corresponds to any agent's (the human and the technology) "program of action." In the words of Latour (32), the term means "displacement, drift, invention, mediation, the creation of a link that did not exist before and that to some degree modifies two elements or agents." The absence of the human agent instead characterizes delegation. Such is, for instance, the case of the speed bump or sleeping policeman, in which the designer has inscribed her program, and that for this reason also works in the complete absence of any actual policeman. In delegation, technology becomes in a certain sense autonomous. In the second part of the book, I will try to go beyond delegation. While the speed bump is still a *passive* technology, digital entities like unsupervised algorithms of machine learning are increasingly *active* and *autonomous* in their decisions. The question I will ask in the second part of the book is if it possible, and eventually to which extent, to attribute to digital machines, or at least to an emerging part of them, interpretational capacities.

I will enter the debate by presenting the anthropocentrism that has classically characterized hermeneutics. I will then follow two different strategies. In chapter 3, I will 'exalt,' so to say, the digital machines' capacities to schematize (that is, to interpret) the world. I will resort to the Kantian notion of productive imagination to argue both that human schematizations are always externalized and materialized into technologies, and that human schematization has its analog in the digital dynamics of articulation between databases and algorithms. In chapter 4, I will 'frustrate' human pretentions in terms of creativity, originality, and authenticity. More specifically, I am going to introduce a distinction between imagination-bricolage and imagination-engineer, and I will advance the thesis according to which despite our claims 'we have never been engineers.' We are in sum more hetero-determined than what we are usually disposed to admit, and we are maybe less different from digital machines than what we believe. In the conclusion of this second part, I will present some attempts to overcome anthropocentrism in hermeneutics, paying particular attention to the field of environmental hermeneutics, with the intention to import such an attitude into the context of digital hermeneutics.

The book ends with a rather long finale. Whereas the first part of the book mainly focuses on epistemological and methodological issues, and the second part on ontological matters, the finale will take an ethical and political stance. It will be the occasion to go beyond the apparent contradiction between the cautious attitude of the first part and the more enthusiastic approach of the second one. There are in fact several intrinsic differences between human and digital schematizations. Furthermore, between them, there is a specific form of interference, a sort of mutual affection, that, I will argue, has the character of indifference. Such indifference has the ruinous effect of anesthetizing our subjective and existential movements. I will use the Bourdieusian notion of *habitus* for arguing that within the digital we are repeatedly being reconducted to an oversimplified image and understanding of ourselves. Before such standardization of our intentions and affections, some answers seem possible. For instance, while digital tools are reproductional machines (*machines à reproduction*), it is also true that the same machines materialize our action and offer in this way the occasion to take a certain hermeneutic distance from our daily behaviors. But the most powerful source of response lies in my opinion in education, and more widely in institutions. For Bourdieu, the educational institutions are the places par excellence of the reproduction of exclusion, discrimination, and more generally of social classes. Yet, for him, these institutions can also have an emancipative role. While in critical theory institution rhymes typically with illusion and domination, I am going to state that institutions can (which, of course, does not mean that they always *are*) also be the bearer of a general project of reconfiguration of the dynamics between the subjects and the sociotechnical systems.

Notes

1. Latour's critique is slightly different from mine. For him (Latour 1993, 63–64), semiotics, semiology and all other linguistic turns "render more difficult the connection between an automatized discourse and what they had provisionally shelved: the referent [. . .] and the speaker." This is, in fact, the limit of structural linguistics and all other approaches privileging synchronicity over diachronicity. But this is not the limit of philosophical hermeneutics, which generally focuses on the relation between meaning and reference, text and context, et cetera.
2. Similarly, for Jacques Ellul (1964, 142) Technology is the adversary (and one should not forget that "adversary" is the meaning of "Satan" in Hebrew) of God. Technology desacralizes the world, and paradoxically, for this reason, is sacralized by human beings. God and Technology contend the same role before human beings.
3. According to Coeckelbergh and Reijers (2016, 328), Kaplan has well analyzed how Ricoeur's work can explicate the ways in which technologies mediate social meanings. However, "he remains silent about what a proposed synthesis of philosophy of technology and Ricoeur's work might actually look like."
4. The fact that both the Bible and Aristotelian works were originally written on scrolls is irrelevant here because Ricoeur uses the *Poetics* and the Bible to account for the printed tradition of modern and contemporary narratives.
5. Against the ethical prescription of being narrative, that is, to see one's own life as a continuum, see Strawson (2004), who considers himself an "episodic" kind of person, and yet has moral values and certain psychological stability.
6. https://ich.unesco.org/en/what-is-intangible-heritage-00003. Accessed June 10, 2019.
7. All translations from French, German, and Italian are mine.

References

Bouchardon, S. 2014. *La valeur heuristique de la littérature numérique*. Paris: Hermann.

Coeckelbergh, M., and Reijers, W. 2016. "Narrative Technologies: A Philosophical Investigation of the Narrative Capacities of Technologies by Using Ricoeur's Narrative Theory." *Human Studies* 39(3): 325–346.

De Mul, J. 2015. "The Game of Life: Narrative and Ludic Identity Formation in Computer Games." In J. Goldstein, and J. Raessens (eds.), *Handbook of Computer Games Studies*. Cambridge, MA: The MIT Press, 251–266.

———. 2010. *Cyberspace Odyssey: Towards a Virtual Ontology and Anthropology*. Newcastle upon Tyne: Cambridge Scholars Publishing, 161–192.

Ellul, J. 1964. *The Technological Society*. New York: Vintage Books.

Gransche, B. 2017. "The Art of Staging Simulations: Mise-en-scène, Social Impact, and Simulation Literacy." In M.M. Resch, A. Kaminski, and P. Gehring (eds.). *The Science and Art of Simulation*. Dordrecht: Springer, 33–50.

Grondin, J. 1993. *L'universalité de l'herméneutique*. Paris: P.U.F.

Heidegger, M. 1996. *Being and Time*. Albany: SUNY Press.

Ihde, D. 2010. *Heidegger's Technologies: Postphenomenological Perspectives*. New York: Fordham University Press.

———. 1998. *Expanding Hermeneutics: Visualism in Science*. Evanston: Northwestern University Press.

Kaplan, D. 2006. "Paul Ricoeur and the Philosophy of Technology." *Journal of French Philosophy* 16(1–2): 42–56.

Latour, B. 2007. "Can We Get Our Materialism Back, Please?" *Isis: A Journal of the History of Science Society* 98(1): 138–142.

———. 1994. "On Technical Mediations." *Common Knowledge* 3(2): 29–64.

———. 1993. *We Have Never Been Modern*. Cambridge, MA: Harvard University Press.

Lewin, D. 2012. "Ricoeur and the Capability of Modern Technology." In T.S. Mei, and D. Lewin (eds.). *From Ricoeur to Action: The Sociopolitical Significance of Ricoeur's Thinking*. London and New York: Continuum, 54–71.

Loute, A. 2017. "L'imagination au cœur de l'économie de l'attention: l'optimisme semantique de Paul Ricœur." *Bullettin d'analyse phénoménologique* 13(2): 494–524.

Michel, J. 2015. *Quand le social vient au sens: Philosophie des sciences historiques et sociales*. Brussel: Peter Lang.

Mitcham, C. 1994. *Thinking Through Technology: The Path Between Engineering and Philosophy*. Chicago: The University of Chicago Press.

Ricoeur, P. 2004a. *Memory, History, Forgetting*. Chicago: The University of Chicago Press.

———. 2004b. *À l'école de la phénoménologie*. Paris: Vrin.

———. 1991. *From Text to Action: Essays in Hermeneutics, II*. Evanston: Northwestern University Press.

Romele, A. 2018. "Herméneutique (du) digital: les limites techniques de l'interprétation." *Études Digitales* 3: 55–74.

———. 2014. "The Ineffectiveness of Hermeneutics. Another Augustine's Legacy in Gadamer." *International Journal of Philosophy and Theology* 75(5): 422–439.

————. 2013a. *L'esperienza del verbum in corde: Ovvero l'ineffettività dell'erme-neutica*. Milan: *Mimesis*.

————. 2013b. "Narrative Identity and Social Networking Sites." *Études Ricœuriennes/Ricoeur Studies* 4(2): 108–122.

Severo, M. 2018. "Safeguarding Without a Record? The Digital Inventories of Intangible Cultural Heritage." In A. Romele, and E. Terrone (eds.). *Towards a Philosophy of Digital Media*. London: Palgrave Macmillan, 165–182.

Stiegler, B. 2014. "Pharmacologie de l'épistèmè numérique." In B. Stiegler (ed.). *Digital Studies: organologie des savoirs et technologies de la connaissance*. Limoges: FYP Éditions, 13–26.

Strawson, G. 2004. "Against Narrativity." *Ratio* 17(4): 428–452.

Verbeek, P-P. 2005. *What Things Do: Philosophical Reflections on Technology, Agency, and Design*. University Park: Penn State University Press.

Wolff, E. 2013. "Compétences et moyens de l'homme capable à la lumière de l'incapacité." *Études Ricœuriennes/Ricoeur Studies* 4(2): 50–63.

Part 1
The Virtual Never Ended

Introduction

As I have said in the overture, among the mediators that exist between the subject and the world, hermeneutics has generally favored the language, while neglecting other mediators, and ignoring the materiality and the technicity of the multiple supports for transmitting meaning—starting from vocal expression (Cavarero 2005).

And yet I do not believe that one should underestimate the importance of language, both in the philosophy of technology and in the philosophy of the digital. Technologies are always embedded into systems of signs and symbols that mediate their understanding and uses, individually and socially. In this case, one can talk of technical or sociotechnical imaginaries such as those analyzed in Jasanoff and Kim (2015). In a recent article, Johnathan Grey presented "data worlds" as horizons of intelligibility. According to him (2018, np), "[j]ust as industrial technologies of the past were accompanied by new social, cultural and political imaginaries, so we can trace the ascent of 'data imaginaries' and 'data speak': visions and rhetoric concerning the role of data in society." Taine Bucher (2017) has introduced the concept of "algorithmic imaginaries," that is, the ways of *talking* about what algorithms are, how they function and how they should be. For the author, such an imaginary has a generative role in molding the algorithms themselves. After more than twenty years of empirical turn in philosophy of technology, it is now time to readmit what has been brutally defenestrated in the past.

But even while remaining within the limits of the empirical turn, as it will mostly be the case in this book, the digital is deeply concerned with language, because in the digital everything (sounds, images, et cetera) is translated into writing or, to be more precise, 'transcoded.' As previously said, the specificity of this writing and its symbols is to be both representative and performative, readable (after a certain amount of training) by a human being and executable by a machine. As the French philosopher and computer scientist Bruno Bachimont has stated (2010, 153), information and communications technologies have determined a hybridization between writing systems (intellectual technologies) and production systems (material technologies): the articulation between an expertise about expression, contents' transmission and their consultation, and an expertise about the transformation of the matter. Cybernetics allowed us to understand the physical systems in terms of information; Hilbert's formalism helped us to grasp expression and representation as techniques of formal manipulation, thus making possible informatics and algorithmics. Incidentally, to readmit the language in philosophy of technology does not mean to return to the idea of a radical difference between signs and technologies, which would be a specific declination of the even older distinction between spirit and matter. The difference between signs and technologies is not ontological, but entirely material and technological: "The symbolic polimorphism is a particular material politechnology" (Hottois 2017, 95). It is precisely for this reason that, I believe, hermeneutics still has a role to play in understanding the digital. As I have already argued, digital technologies are hermeneutic in the sense of the hermeneutic relations described by Don Ihde (1990, 80–97), in which the technology offers a representation of the world that must be interpreted to access the world. This is the case, for example, with maps, thermometers, and flight instruments. This also holds true of digital technologies, which offer representations of the world that are interpreted by a human, a non-human, or, as it is most often the case, by a combination of both.

However, digital technologies are hermeneutic technologies of a specific kind since they seem to have the capacity of minimizing the distance between the world and its representations. To put it differently, the digital might be seen as the technological ensemble that has the power (and the pretention) to undo the gap between the "map" and the "territory." I am referring to the famous short text by Jorge Luis Borges "On the Exactitude of Science," attributed to the fictional author Suárez Miranda:

> In that empire, the art of cartography attained such perfection that the map of a single province occupied the entirety of a city, and the map of the empire, the entirety of a province. In time, those unconscionable maps no longer satisfied, and the cartographers guilds struck a map of the empire whose size was that of the empire, and which coincided point for point with it. The following generations, who were not so fond of the study of cartography as their forebears had been, saw that that vast map was useless, and not without some pitilessness was it, that they delivered it up to the inclemencies of sun and winters. In the deserts of the West, still today, there are tattered ruins of that map, inhabited by animals and beggars; in all the land there is no other relic of the disciplines of geography.[1]

Less famous is probably a text by Umberto Eco (1994), in which he takes seriously Borges's story. For him, to have a map of the empire on a scale 1 to 1, it would be necessary that the map (1) not be transparent, or (2) not lie on the territory, or (3) be adjustable in such a way that the reference points of the map lie on points of the territory that are not the ones they indicate. But each of these three conditions involves insuperable practical difficulties and theoretical paradoxes. For instance, from a practical point of view, he shows the limits of different solutions such as an "opaque map spread out over the territory," a "suspended map," and a "transparent map, permeable, extended, and adjustable."

The digital seems precisely to be able to overcome most of these difficulties, firstly, because there is a high degree of dematerialization. Although the digital is undoubtedly not merely spiritual, it is evident that its material substructures or infrastructures are less bulky than those that would be necessary to have the same results on printed paper. To use one of Eco's images, one could say that for the digital there is still a vast desert, whose limits have not been explored or seen yet, where the map can be fold and unfold at will. Secondly, there is the high information resolution. As I will argue in the following pages, what most characterizes the digital today is not information and communication, but registration and recording. Digital traceability has become "a total social fact." There is no aspect of reality that cannot not be digitalized; this is also because the digital is pure calculation and manipulation. According to Bachimont (2010, 155–156), the digital has two main properties: (1) a twofold independence of its signs from sense and meaning—they are defined independently of each other, and they have no specific signification, and (2) the fact that the rules of manipulation of signs are formal and mechanical. Recently, he has also spoken of computer science as "a spiritual asceticism of meaning": "A difficult exercise for us human beings, who are above all semiotic animals that approach what surrounds us by its significance, as a message that we must interpret since the world is not merely reduced to what is shown here and now" (23). Thirdly, one could argue that no matter how big the map is, our perceptive and cognitive capacities are limited. And yet, the digital offers also several instruments to render such vastness and complexity graspable for us, for instance, data visualization techniques. Also, web browsers, news aggregators, et cetera, are tools the digital offers to make itself more bearable to us.

The thesis I would like to defend in this part of the book is that the frontiers between the territory and the map, between the reality and its digital representations, are now more porous, but they exist and still resist. In other words, my thesis is that 'the virtual never ended,' an expression which is a tribute to Philip K. Dick and his idea that the Roman Empire never came to an end. Between 1974 until his death in 1982, the American writer kept a journal, the *Exegesis*, in which he documented and reflected on his religious and visionary experiences (a volume of excerpts of 944 pages has been published in 2011). One of the recurrent themes of *Exegesis* (which is also at the center of Dick's novel *VALIS*) is that history has stopped in the first century CE and that "the Empire never ended."

The Roman empire is keeping the population enslaved, and its apparent demise plays a fundamental role in its continuous flourishing.

The notions of virtual and virtuality used to indicate a "spaceless space," that is a dimension separated from real life and its physical and social constraints. According to several digital pioneers, people could experiment with new interactions and configurations of oneself in the virtual without much risk. Such experimentations could even have empowering effects "IRL, in real life."[2] Today, nobody in the field of digital studies would dare to resort to this old-fashioned concept. And yet, what if the virtual was still there and most scholars were duped, victims of an illusion or, to be less drastic, of a dominant digital worldview? And what if part of the power and fascination of the digital was precisely coming from this concealment?

Hermeneutics and, more specifically, the hermeneutic dialectics between distanciation and appropriation can be used just to defend this thesis. The problem is that hermeneutics is not particularly liked by many: too 'soft' to be admitted among the philosophies of substance, but still too 'hard' to be part of the philosophies of becoming. It is precisely because of its mixed nature that hermeneutics has not been able to find space of its own in the contemporary debate on the digital, with a few exceptions that I am going to discuss. Indeed, if the Internet, the digital technology par excellence, is a network of networks, why should one use a theory that has undoubtedly given importance to relations, though not *enough*? It is easy to understand why today one prefers to approach the digital through postmodern thinking rather than through authors such as Ricoeur, Gadamer, or Heidegger—who also accused, in his 1962 conference "Traditional Language and Technological Language" (1998), information cybernetics to be the most violent and dangerous aggression against the *logos*. It is also understandable that one favors fluid conceptualities rather than the mechanical terminology of hermeneutics, because eventually the hermeneutic circle remains a gearing, entangling as much as maintaining distinctions.

As I have said in the overture, while Gadamer mostly perceived distanciation negatively, as a form of alienation and estrangement from the object of interpretation, Ricoeur understood it positively, as a counterbalance to the too quick appropriation and "fusion of horizons." For the French philosopher, distanciation is the middle term between an immediate and a more reasoned form of appropriation. On the one hand, distanciation indicates the 'cold' (or let's say 'scientific' and 'hard') methodologies that an interpreter might use to approach her research object. On the other hand, it is used to stress the independence of this object, mostly a text, from the intentions of its author and reader, observer, et cetera. While referring to methodologies for understanding texts, Ricoeur was thinking of semiotics and linguistic structuralism. Today, digital humanities and methods have a prominent role. But what interests me most here is that digital technologies are always based on a process of symbolic distanciation from the world, and it is only on this basis that they appropriate the entire world and become effective into it. Those who believe in 'the end of the virtual' are victims of an illusion of transparency. And it is precisely the often-invisible hermeneutic dynamics of distanciation and appropriation of the digital that makes room for interpretation in the digital. In other words, it is because the digital is structurally based on a performative distanciation from the world that the digital cannot realize its aspiration of being the world, but rather continues to need the world as its interlocutor and as its 'otherness.'

Notes

1. https://en.wikipedia.org/wiki/On_Exactitude_in_Science. Accessed June 10, 2019.
2. In the literature dedicated to online environments between the 1980s and 1990s, the term "virtual" indicated, more or less implicitly, three things at once: firstly, a "spaceless space," as Manuel Castells defined it. Secondly, the virtual was viewed as an opportunity to experience new possibilities, options, and actions without the risks of "true life." This second meaning is closer to the etymology of the word. Indeed, "virtual" comes from the Latin *virtus/virtualis*, a direct translation of the Greek term *dynamis*, which can be transcribed as "ability," "potentiality," or "power." In his commentary to the beginning of the ninth book of Aristotle's *Metaphysics*, Heidegger (1995) translated *dynamis* with *Kraft*, "force" in English, as well as *Vermögen*, a word which means "ability" but also "capacity" and "capability." As such, the virtual is not opposed to the real, but rather to the actual—*actus* is the Latin translation of *energeia*. And this is precisely the third meaning of the term: virtual as individual and social empowerment.

References

———. 2010. *Le sens de la technique: le numérique et le calcul*. Paris: Les Belles Lettres.

Bucher. T. 2017. "The Algorithmic Imaginary: Exploring the Ordinary Affects of Facebook Algorithm." *Information, Communication & Society* 20(1): 30–44.

Cavarero, A. 2005. *For More than One Voice: Toward a Philosophy of Vocal Expression*. Stanford: Stanford University Press.

Eco, U. 1994. "On the Impossibility of Drawing a Map of the Empire on a Scale 1 to 1." In *How to Travel with a Salmon and Other Essays*. San Diego: Harcourt, 95–106.

Grey, J. 2018. "Three Aspects of Data Worlds." *Krisis: Journal for Contemporary Philosophy*. http://krisis.eu/three-aspects-of-data-worlds/. Accessed June 1, 2019.

Heidegger, M. 1998. "Traditional Language and Technological Language." *Journal of Philosophical Research* 23: 129–145.

———. 1995. *Aristotle's Metaphysics 9, 1–3*. Bloomington: Indiana University Press.

Hottois, G. 2017. *Philosophie et ideologies trans/posthumanistes*. Paris: Vrin.

Ihde, D. 1990. *Technology and the Lifeworld: From Garden to Earth*. Blooming-ton: Indiana University Press.

Jasanoff, S., and Kim, S-H. 2015. *Dreamscapes of Modernity: Sociotechnical Imaginaries and the Fabrication of Power*. Chicago: The University of Chicago Press.

1 The Virtual Invaded the Real

In this first chapter, I will take a position against those who more or less explicitly defended the idea according to which 'the virtual invaded the real.' I refer to the theorists who somehow believe in the total digitalization of our human condition in the present or next future. I could have criticized transhumanism. I could have chosen, for instance, the easy path of an analysis of the problematic notion of singularity in its different meanings. But I have decided to tread a harder path and to deal with Luciano Floridi's semantic theory of information (STI). In some sense, my approach to the digital wants to be a hermeneutic reply to such a theory. In other words, my approach is the *communicational* counterpart of the *informational* approach to the digital. Although it is certainly more complex and philosophically convincing, I have the impression that some ethical and ontological consequences of Floridi's theory of information entertain more than a few resemblances with the transhumanist suppositions about the present situation and the destiny of the humankind. I would like to introduce now the debate by making a few remarks on the notion of information in general to justify the importance I attribute to Floridi's approach in this context.

The term "information" derives from the Latin *informatio*, which has two meanings: "the action of giving shape to something material" and "the action of communicating knowledge to another person." The prefix *in-* indicates the fact that we are dealing with an action that consists in giving form or knowing it and/or making it known. The link with the Platonic and above all Aristotelian notion of form, which has both an ontological and epistemological significance, is evident. Capurro (2009, 128) states that the Latin term has the same or similar meaning to the Greek words *idea*, *eidos*, *morphe*, or *prolepsis* in authors such as Cicero, Augustine, and Aquinas. For example, in the book XI (2, 3) of *De Trinitate*, Augustine calls the process of visual perception "information of the sense (*informatio sensu*)," referring to the Aristotelian image of memory as a wax table on which forms are impressed. In Aquinas, then, Aristotelian hylomorphism becomes the process of "information of (or giving shape to) matter" (*informatio materiae*). Capurro (129) writes that the transition from the Middle Ages to modernity marks the shift from a substantial—"giving a (substantial) form to matter"—to a subjective conception of information—"communicating something (new) to someone"—and that this passage is evident above all in Descartes. The author is inspired by Peters (1988, 12), for whom "in the feverish demolition of medieval institutions in the seventeenth and eighteenth centuries, the notion that information consisted in the activity or process of endowing some material entity with form remained largely unchanged." And yet, for Peters, "the notion that the universe was ordered by forms fell into disrepute, and the context of this informing shifted from matter to mind." There has been, thus, a radical inversion in the meaning of information.

As Adriaans writes in the entry "Information" of the *Stanford Encyclopedia of Philosophy* (2012, np) in the *Meditations* the wax metaphor, which for over fifteen hundred years had been used to explain the sensory impression, "is used to argue *against* the possibility to reach knowledge via the senses. Since the essence of the *res extensa* is extension, thinking fundamentally cannot be understood as a spatial process." The price to pay for this inversion is high, too high because ideas, forms, and, therefore, the fact of "informing" and "informing oneself," once rendered completely independent from the senses, will tend to become innate and a priori. For this reason, Adriaans (np) maintains that Locke's reinterpretation of the notion of idea as a "structural placeholder" for every entity present in mind is an essential step toward the emergence of a truly modern notion of information. In fact, with Locke and then with Hume we move on to a completely probabilistic and a posteriori form of knowledge: "The construction of concepts on the basis of a *collection of elementary ideas* based in sensorial experience opens the gate to a reconstruction of *knowledge as an extensive property of an agent*: more ideas implies more probable knowledge" (np). The author refers to the formal theories of probability developed by authors like Pascal, Fermat, and Huygens.

Precisely in a probabilistic way Claude Shannon defined information in the twentieth century.[1] Of the old meaning of information, in which it is, first of all, a matter of giving or transmitting the form to things, something survives in Gilbert Simondon, who, among other things, was one of the organizers of the sixth symposium at Royaumont in 1962 to which Norbert Wiener and Benoit Mandelbrot participated. In Simondon, one can speak of a real 'informational ontology,' in the sense that he thinks of information as a substitute for the notion of form, which does not refer so much to the fact of giving form to an informal subject, as to a process of individuation or even better of concretization that involves the whole of the entity in question. Noticing some similarities with the philosophy of Luciano Floridi, Andrew Iliadis (2013, 18) writes that "a Simondonian information ontology finally authorizes us to put aside the deadlock of the subject-object distinction and instead consider the human being present in the technological object and vice versa, as a whole." According to me, the fundamental difference between Floridi and Simondon lies in the attention given by the latter to the notion of "transduction," which has both epistemological and ontological value, and which serves to detect both similarities and differences between different modes of existence. I will discuss this point below.

To return to Shannon, it is known that his definition of information is strictly syntactic. As he writes at the beginning of his article "A Mathematical Theory of Information" (1948), "Frequently the messages have *meaning*; that is, they refer to or are correlated according to some system with certain physical or conceptual entities. These semantic aspects of communication are irrelevant to the engineering problem." And it is precisely this suspension of meaning and reference that allowed Shannon first and then Wiener to propose elegant mathematical definitions of information. Both Wiener and Shannon tie the concept of information to the notion of entropy in statistical mechanics. The former has defined the information in terms of negative entropy: "Just as the amount of information in a system is a measure of its degree of organization, so the entropy of a system is a measure of its degree of disorganization; and the one is simply the negative of the other" (Wiener 1965, 11). Shannon, on the other hand, defines it in terms of positive entropy. De Mul (1999, 83) writes in this regard:

> For Shannon, the informational value is defined as the measure of freedom with regard to the selection of elements which exists in a communication process. The greater the freedom of choice, the greater the uncertainty and therefore the greater the entropy. In this sense, a page containing letters placed at random has greater informational value than a page of Kant's *Critique of Pure Reason*.

There is undoubtedly a philosophical richness already inherent in the mathematical theory of communication. Just think of the themes of probability and of the "model selection" that have generated interesting philosophical reflections as in Adriaans and Vitanyi (2009) the last few years. But I believe that hermeneutics enters into play in this context when the questions of meaning, reference, and truth arise. And this is the reason why I consider Floridi's approach to information particularly relevant, since he tried to give a definitive answer to such questions. The goal of this section consists of showing the implausibility of Floridi's solution, and into paving the way for an alternative approach.

Floridi is drastic, but he is for sure not wrong, arguing that Shannon's theory is not even an information theory, but just a "mathematical theory of data communication." In his reinterpretation of Shannon's perspective,

Weaver (Shannon and Weaver 1964, 4), on the other hand, immediately distinguishes between three levels of the communication problem: (1) how carefully the symbols of communication can be transmitted—technical problem; (2) how precisely the transmitted symbols express the desired meaning—semantic problem; (3) how effectively the meaning received affects the conduct in the desired manner—problem of effectiveness. Incidentally, the topic of effectiveness is that of Wiener's theory of cybernetics, which is about the relationship between communication and control: "When I communicate with another person, I impart a message. [. . .] When I control the actions of another person, I send you a message" (Wiener 1989, 16). This threefold distinction is not without analogies with the distinction proposed by Charles Morris between a syntactics, a semantics, and a pragmatics of the sign. The first concerns the formal relations between signs; the second is about the function and the meaning attributed to the signs; the third refers to the relationship between the signs and their user and receiver.[2]

The thesis that I want to advance in this context is that there are two ways of understanding the relations between information and meaning (where, in the proposed perspective, meaning implies at the same time semantics and pragmatics, since there is no meaning that makes sense out of a specific context of use). The first one is the one theorized by the Oxford philosopher, who makes this relationship internal and even essential, in the sense that for him there is no information that does not already contain meaning and that is not truthful. The second one, which instead represents a hermeneutic theory of information, understands meaning and truth as secondary elements, not in the sense that they are less relevant, but in the sense that they can be attributed only afterward.

The reflection will be developed in three steps. Firstly (1.1), I will discuss Floridi's STI. I will particularly insist on the problematic idea according to which truth must be from the beginning 'encapsulated' in the semantic information. Secondly (1.2), I will account for some ethical, ontological, and technological consequences of this theory. I am going to highlight, for instance, the irreconcilability between Floridi's macroethics and his notion of "levels of abstraction." Thirdly (1.3), I will propose an alternative theory of information that I call hermeneutic. Since I have no ambition in becoming a Floridi scholar, neither do I have the skills to understand the technical logic formulas he abundantly resorts to in publications like Floridi (2011), I will limit my considerations to a few publications of his, which I consider representative of his vast production. I assume that my strategic reading might sound here and there unfair before a work that deserves (and has already given rise to) more detailed reflections, but I am indeed not the first nor the last to betray the intentions of an author for indulging one's own interests.

1.1 The Semantic Theory of Information (STI)

The definition that Floridi proposes for semantic information can be summarized in the formula: data + meaning + truth.

As for data + meaning, he affirms that this is the definition of DOS, the declarative, objective, and semantic information, which has established itself both in popular culture and in scientific literature, and especially in those contexts dealing with data and information as reified entities: information sciences, information systems theory and management, methodology, database design, decision theory, et cetera (Floridi 2005, 353). The author affirms to be in favor of a "localist principle," for an analysis situated in specific contexts of application. He bases his theory on the dominant idea among the experts of the sectors mentioned above, according to which information is a non-mental, user-independent, declarative, and embedded entity in physical objects such as databases, encyclopedias, websites, and so on. However, one should ask here how legitimate it is to construct a whole theory on the common understanding of those who already belong, with little possibility of real distanciation, to a certain disciplinary area. Would it not be the task of a (critical) philosophy of information to question, rather than embrace, this 'common sense,' which, in fact, sounds like the oldest claim of apodicticity that characterizes the 'exact' sciences? Moreover, how is it possible that Floridi does not say anything about the term "embedded," which refers to a materiality that cannot be taken for granted by a philosopher of technology since each materiality has its affordances? His analysis of the standard definition of information (SDI), which he considers a more rigorous formulation of the DOS, starts not from a problematizing of the materiality and the dependence of the information from its context of production and understanding, but from a suspension of its semiotic code and its physical implementation. In a sort of vicious circularity, the author can only arrive at conclusions that are already implicit in the starting conditions of his theory.

As for SDI, Floridi states that (1) it must consist of n data; (2) data must be well-formed (they must have a syntax); and (3) well-formed data must have meaning (they must respect the meanings of the chosen system, code, or language in question).[3] The author then develops three reflections that lead to three rules of neutrality.[4] First of all (354–359), he underlines that according to SDI, information cannot be dataless, but that data must generally be understood as relational entities, as lack of uniformity—Taxonomic Neutrality, TaxN. This is what Floridi (2010, 23) also calls "diaphorical interpretation" of data since *diaphora* is the Greek word for "difference." Secondly, he speaks of Ontological Neutrality, ON, in the sense that there can be no information without representation, but ON must not be materialistically understood.

These first two rules represent the suspension of the critical judgment about the dependence on context and materiality respectively. As far as the relational nature of data is concerned, Floridi limits himself to briefly underline that data as relational entities need queries to become information: "data are definable as constraining affordances, exploitable by a system as input of adequate queries that correctly semanticize them to produce information as output" (Floridi 2005, 357). But what are these "queries" if not the context within which data as differential entities can and must be understood? Does it make sense to talk about data outside their context of use and interest that, in the case of academic research, would be made of the questions and concepts that already guide it? As for the ON, the author refers to the thesis of authors such as Berkeley and Spinoza, in which all entities, properties, and processes are ultimately noetic—as if to say that ON can be understood as an immaterial thesis within a system that already hypothesizes an essence or a spiritual principle at the heart of the whole reality. Translated into the language of the Oxford philosopher that will be introduced later, the ON is an immaterial thesis within what is already an informational ontology.

Finally, Floridi also presents what he calls Genetic Neutrality, GN, according to which information can have a semantics independently of any informee. The example he proposes is particularly indicative. He writes that "before the discovery of the Rosetta Stone, Egyptian hieroglyphics were already regarded as information, even if their semantics was beyond the comprehension of any interpreter. The discovery of an interface between Greek and Egyptian did not affect the hieroglyphics' embedded semantics but its accessibility" (359). One can deduce that according to Floridi, semantics is in some sense and to a certain extent a property intrinsic to information itself.

But (1) all the observers could have been wrong. They could have believed to be confronted with a form of writing endowed with syntax and meaning when this was not the case. Before the discovery of the Rosetta Stone, many people thought that hieroglyphics were merely images, semantics without syntax, icons and not symbols in Peirce's terminology. And think of those signs like the stains on the walls or the shapes in the sky that are willingly confused, even today, for divine or alien messages; (2) even the mere intuition of a possible syntactics and semantics requires a human intentionality, or at least an intentionality capable of entering a relationship with the human way of giving meaning to things. A dog, on the Rosetta Stone, would have directly urinated, which, in its way, however, is a manner of giving it a semantic function, though different from ours. In this case, following Peirce, we would not speak of symbols or icons but of indexes, characterized by physical contiguity between the sign and the thing to which the sign refers.

It is also interesting to observe how Floridi, despite this position, refuses to assimilate his information theory to an environmental theory of information.⁵ But I do not understand what kind of information theory is a theory that, on the one hand, rejects, for information, both the dependence on the context (or that at least already lies beyond it) and, on the other hand, its environmental dimension. Of course, one could say that the author uses strategically here an example that concerns human beings and not stones, plants, or animals. But to this objection we could answer that, even if distanced over time, in the case of the Rosetta Stone, it is precisely a matter of an encounter between two (or more) human intentions, and is therefore in no case information that has meaning in itself. Moreover, the work of translation, once the interface has been found, could be understood as a work of reconstruction, according to one's own syntactic and semantic framework, of that original intentional process that allowed to give meaning to signs that, in themselves, do not say anything.

Now we come to the complete formula of Floridi's theory: information = data + meaning + truth. According to the Oxford philosopher, SDI defends what he calls alethic neutrality, namely the fact that alethic values are not embedded in but supervene on semantic information. Strictly speaking, "meaningful and well-formed data qualify as information, no matter whether they represent or convey a truth or a falsehood or have no alethic value at all." The most important consequence of this perspective is that "false information (including contradictions), i.e., misinformation, is a genuine type of DOS information, not pseudo-information" (359).

The philosopher formulates then a series of arguments precisely against this idea. Of these, only the third is held here: to the statement that "FI [false information] can still be genuinely informative, if only indirectly," he replies that "this is vague, but it can be reduced to the precise concept of non-primary information µ discussed in section two" (361). Non-primary information is presented through examples in section 2 of the same article to show that although it is true that there is no information without data, data must be understood as differential entities, similar to a black dot standing out against the background of a white sheet. A machine that always gives the same answer does not provide any information unless it stops responding, but the information it will give in this case (for example, the machine is broken) will be meta-information. Another example is that of the wife who does not hear her husband's questions. Silence can be understood as information—a silence that counts as assent as if it were the 0 of a binary code in which the word is 1—or as meta-information: silence is an indication that the wife has not heard the question.

For me, what is problematic in Floridi's argumentation is the "meta-" location of this kind of information. He is certainly right to argue that an interpretative effort is needed to capture the information of a broken machine or a deaf wife. In fact, it is a matter of making a double movement, and collocating the fact, the thing that has been said/heard or not said/heard, within a broader or otherwise different context, so that it can make sense for me again. This effort does not seem essentially different from what I do in situations where things go without saying. In both cases, it is a matter of contextualizing, a fact or an instruction. The difference is that sometimes things go out of habit, because the experience has been reiterated a sufficient number of times and has been incorporated. In other situations, on the other hand, it is a matter of making a new interpretative effort. I would speak in the first case of "white" or "dead" interpretations, and in the second of "living" interpretations.[6]

Take the case of the dialogue between two builders presented by Wittgenstein at the beginning of the *Philosophical Investigations*. The meaning of that example is to show how language—and, for us, we could say semantic information—always depends on a context of use. Now, imagine that the builder B, the one who receives the orders, has to work with a builder A, who is a foreigner and who misuses some terms compared to the usual canons dictated by the language and its context of use. Of course, it is possible to imagine builder B repeatedly correcting his boss and bringing him back to the standard use of words, but it is more likely that he learns to recontextualize the false information given by builder A to make it work in the context in which they both are, without offending anyone. And it is equally probable that the repeated hearing of a wrong word or order, but a correct practice, leads him to develop a new habit so that he might find himself mistaken if one day he finally received the accurate information or order. The same applies to someone dealing with machines that break easily or with an often-distracted wife. Just think of all the situations in which, instead of repairing something, we adapt to its new way of functioning, and we do so if the situation seems to us, so to speak, bearable.

Incidentally, the example of the dialogue between builders could also be used to refute the clear-cut distinction proposed by Floridi between factual and instructional information. There is no information that does not have validity within a specific context of use. Even many perceptual data turn out to be dependent on an extra-perceptive, linguistic, symbolic, cultural, and pragmatic (in the end, hermeneutic) context. This does not mean embracing a relativistic principle, since, on the one hand, things have their affordances, and, on the other hand, culture, language, and so on represent frameworks within which information assumes a defined truth value. Floridi prefers factual semantic information (Floridi 2010, 49–51) only because the instructional one would pose many problems for his rigid theory. For example, one does not understand his thesis according to which "instructional information does not qualify alethically (cannot be correctly qualified as true or false)" (Floridi 2015, np), since the truth value is merely displaced from the sentence to the execution (or not) of that instruction.

In short, one could say that Floridi is undoubtedly right in saying that semantic information needs truth, but that it is wrong in not recognizing the process of attributing meaning and truth which, in varying degrees, is always put in place when semantic information appears. One can then share what is said by Adriaans (2010, 43), according to which "[Floridi's] semantic theory of information prima facie cannot explain this subjective [I would rather say embodied, social, cultural, et cetera] relational aspect of information. It is bound to make information a monolithic static notion that exists independent from any individual observer"; and again: "semantics is not something that stands aside from information theory and can be plugged in at will, as happens in Floridi's theory of semantic information. Theory of information inherently implies a treatment of [and a concern for] the notion of meaning" (52). For Adriaans, this can be done by replacing the problem of founding knowledge as a true and justified belief with that of selecting the optimal model that best fits with specific observations—and, one might add, for a particular purpose and observer-dependent situation. If in Floridi's theory meaning and truth are 'encapsulated' in the very notion of information, it is instead, for me, a matter of 'decapsulating' them. This, of course, does not mean renouncing all concerns about meaning and truth—it would, in fact, be a merely syntactic or mathematical theory of information—but questioning the way in which meaning and truth become part of the information on a case-by-case basis, a sort of third way, then, between those who rule out the problem and those who instead, like Floridi, take it for granted.

Finally, it must be admitted that Floridi himself (2015, np) seems to have moderated his perspective, stating that since the data are at the same time the resources and the limits that make the construction of information possible, the information can also be understood as the result of a data modeling. This perfectly meets what has just been said, namely, that information, its meaning, and its truth are constructed starting from the affordances of data within a (not infinite) multiplicity of contexts of use, which in the specific case of the academic research are the methods, the techniques, and the legitimate questions that guide the research. The notion of Levels of Abstraction, LoA, which plays a vital role in several of Floridi's reflections (2008a), seems, moreover, to be closer to the hermeneutic perspective proposed here. Straightforwardly, one could say that the LoA have to do with the multiplicity of perspectives that different observers have on the same object. For example, a car will be 'seen' differently by a driver, a mechanic, and an insurer.[7] The problem of the LoA theory is not intrinsic to the notion itself. I will use it myself in the second part of the book. The issue instead concerns the possibility of articulating it with his theory of information, which seems non-relativistic (or better, non-pluralist) by nature. There is, of course, the possibility to understand Floridi's theory of information, with its ethical, ontological, and technological extensions, as just one specific LoA from which to observe and understand the new digital world. But, then, his claim of truth about semantic information as inherently truthful would be just a truth among other possible truths.

1.2 Ethical, Ontological, and Technological Consequences of the STI

The thesis implicit in what follows is that there are several "family like-nesses" between the epistemological monolithism just described and criti-cized and those that, in Floridi, could be called an ethical-ontological and a technological monolithism. To put it a bit brutally, the fact of including meaning and truth in the definition of semantic information, making it in some way independent of any exteriority, brings to a night in which all cows are informational.

As Charles Ess writes (2009, 160), Floridi's information philosophy is first and foremost an ontology that takes information as the first among the ontological categories and as the ultimate constituent of reality itself—so much so that the Oxford philosopher (Floridi 2008b, 199) comes to write that "to be is to be an informational entity." The thesis is strong but undoubtedly interesting, with immediate ethical consequences: "To be sure, his ontology resonates with feminist and environmentalist views in critical ways, as well as with Buddhist and Confucian views. But Floridi's PI [Philosophy of Information] is more radical than at least its closest Western cousins, as it starts with the (argued) claim that *everything* is fundamentally information" (Ess 2009, 160).

For example, an essential element is the predilection of the relation (information is always a form of transmission and therefore of communica-tion) and of the multiplicity over unity. In a certain sense, this approaches Floridi's perspective to Latour's actor-network theory, as noted by Adam (2008), but also to Daniel Dennet's "as if" intentionality—Grodinsky, Miller, and Wolf (2008), with which Floridi declares to be in full agree-ment, speak of "intentionality*"—or to the notion of moral mediators developed by Peter-Paul Verbeek and other post-phenomenologists. One can also notice similarities with the relational ontology of Gilbert Simon-don. In this regard, the French philosopher writes that "a relation must be understood as a relation in the being, relation of being, way of being and not simply relation between two terms that we could adequately know through concepts because they have an existence that is separate" (Simondon 2005, 32). In the same text, just a few pages before, Simon-don defines as an individual everything that can preserve and increase its information content to some extent (28).

About ethics (but ethics and ontology, in Floridi, are a one and only thing, in the form of an ethical ontologism), Ess (2009, 161) speaks of "philosophical naturalism," as far as Floridi takes reality qua informa-tion as intrinsically valuable. The informational macroethics proposed by the philosopher of Oxford is an adaptation and even a radicalization of environmental ethics, which emphasizes the intrinsic value of life and the intrinsically negative value of suffering (Floridi 2010, 111). It is a biocentric and patient-oriented approach, in which, however "patient" is not only a human being but every form of life. Floridi proposes to replace "life" with "existence" (from biocentrism to inclusiveness, because there is something that is even more elementary than life, which is being), and "suffering" with "entropy," in the sense of "any kind of *destruction, corruption, pollution*, and *depletion* of informational objects, [. . .] that is, any form of impoverishment of reality" (112). This perspective, firstly, suggests that everything that is "being/information" has an intrinsic value; secondly, it defends the idea that every informational entity has the right to persist, and even a right to flourish in its state.

The consequences of this approach are essential according to Floridi because bioethics and environmental ethics "fail to achieve a level of complete impartiality because they are still biased against what is inanimate, lifeless, intangible, or abstract" (116). But, in fact, even the problems are not few. The informational nature of all reality, and therefore its integral defense, renders some choices that according to a daily and less ontologized ethics would be rather obviously more difficult, such as to distinguish and prefer a human being over, for example, a web bot. Indeed, one could say that, according to an informative perspective à la Floridi, a data processing center is to be preferred, due to its abundance of information, to a single human being and perhaps even to an entire community, which could therefore be sacrificed. Even more paradoxical is the situation when one must deal with people who for their whole lives or a part of them, such as childhood, illness, and old age, find themselves in an information deficit situation. But even in wanting to be charitable toward Floridi's theoretical intentions by believing that he reasonably does not prefer a server to an informationally disabled person, one can say that he forgets that the real ethical problems start not when it is about giving a general right to flourishing to all beings, but when it comes to having to decide if and how to favor some of them at the expense of others. Alas, flourishing is most of the time a rivalrous good. Recognizing to all animals, plants, and rocks their right to flourish is undoubtedly something beautiful and praiseworthy, but not very interesting because it does not solve most problems. How can this perspective be integrated, for instance, with issues concerning veganism, pharmaceutical experimentation on animals, or merely the construction of a new tunnel through the mountains? How can such macroethics say something meaningful about the conflict of interpretations beyond the phenomenon of mass immigration, when different poverties and difficulties (the ones of the immigrants and the ones of the inhabitants of that specific country) are confronted?

Brey (2008, 112) rightly observes that "IE [Floridi's Information Ethics] is committed to an untenable egalitarianism in the valuation of information objects. Within IE, it seems, no difference in value exists between different kinds of information objects. [. . .] [A] work of Shakespeare is as valuable as a piece of pulp fiction, and a human being as valuable as a vat of toxic waste." Floridi's reply (2008b, 190) to this criticism is perhaps even more interesting in this regard because it involves the notion of LoA, that partial relativism and that dependence on the context that seems, in principle, to be excluded from its epistemology, ontology, and ethics in the strict sense. The only way to make Floridi's macroethics work would, then, be to understand it as a "horizon of meaning" to be kept in mind where choices are made within a specific LoA. In this case, however, one wonders if macroethics is still macro-, or if it is not, in fact, subject to microconditions and to commonsense issues on which its possible adoption always depends.

Floridi's idea on the digital are in continuity with his ethical-ontological perspectives. He introduces, among other things, the notion of a "fourth revolution" (Floridi 2014). The Copernican revolution has taught humans that they are not at the center of the universe; Darwin placed them within the animal kingdom, to which they did not think they belonged; the Freudian revolution made them understand that they are not transparent to themselves, nor masters in their own homes, contrary to what Descartes thought. However, from the 1950s, computer science profoundly changed our understanding of both the world and ourselves: "In many respects, we are not standalone entities, but rather interconnected informational organisms or *inforgs*, sharing with biological agents and engineered artefacts a global environment ultimately made of information, the infosphere" (Floridi 2010, 9). The term "infosphere" can be understood in two ways. Firstly, it may indicate the field of the technical production of meaning: "*Minimally*, infosphere denotes the whole informational environment constituted by all informational entities. [. . .] It is an environment comparable to, but different from, cyberspace, which is only one of its sub-regions" (41). Secondly, the infosphere coincides with everything that is informational in nature: "*Maximally*, infosphere is a concept that can be also used synonymous with reality, once we interpret it informationally" (Floridi 2014, 41). The digital is what makes the two definitions collapse, in favor of a new way for humans to inhabit the world: "we are probably the last generation to experiment a clear difference between online and offline environments. Some people already spend most of their time onlife" (92). Thinks to the complex rituality that characterized the connection to the Internet until ten or fifteen years ago: go to the university's computer room, and wait for a free place, sit on a chair and turn the computer on, wait, turn the modem on, wait (with the typical dial-up noise), launch the browser, wait, and so on. Such rituality used to sanction the difference between reality and virtuality, and the passage from the former to the latter. And think how everything became so immediate, smooth, and transparent today.

The *Onlife Manifesto*, of which Floridi is the editor, begins by noting "the blurring of the distinction between reality and virtuality," along with a strong statement about "the blurring of the distinctions between human, machine and nature."[8] We live in an era where, thanks to digital technologies, in particular, the whole world seems to us to be increasingly "ready-to-hand." Our intentions are more easily satisfied; our actions are more effective; we feel more comfortable because the world finally seems to better coincide with our expectations and intentions. We are therefore in the process of bridging the gap between phenomenon and noumenon, will and reality, epistemology and ontology. A new era of homology between thought and being, not just thought but concretely realized, is soon to come. The virtual has invaded the real, and this event has already brought to a "reontoligization of the world" (Floridi 2010, 11–12).

In a fascinating (although, in my opinion, deeply wrong) book entitled *Informatique céleste* (Celestic computer science) (2017), Mark Alizart has carried to extremes a pan-informational (and pan-computational) perspective in some respects similar to that of Floridi, curiously relying on Hegel. According to him, Hegel discovered "before Turing the continuous nature of the world: the ultimate determination of everything is the unity of thought and Being, that is what we precisely call today 'information'" (74). As he writes in the concluding part of his work, "everything is information, the unity of thought and being is the Whole. [. . .] Only to informatics belongs the task of realizing the eschatological promise common to the great utopias and the great religions: that of reconciliation between words and things, between the dead and the living, between humans and non-humans" (190–191).[9] It seems to me that there is something profoundly Hegelian also in Floridi, who paraphrases Hegel when he writes (Floridi 2014, 41) that "what is real is informational and what is informational is real." One could even say that in Floridi there is more Schelling than Hegel since for him nothing is to be conquered and everything is already spirit, since the difference between the real and the virtual, or between matter and information, is always-already outdated. In a sense, it might be said that for Floridi the digital does not transform reality but opens the eyes on what already is. I was not surprised, then, when I first read on the *New York Review of Books* a review by John Searle in which Floridi's *The Fourth Revolution* and Nick Bostrom's *Superintelligence* were criticized together.[10]

1.3 Philosophy of Information: A Hermeneutic Approach

As already mentioned, it is here a question of exploring a "third way" between those information theories that exclude, in principle, any semantic and pragmatic implication and those that instead include them from the beginning. In other words, I want to present an approach to information that focuses on the issue of the attribution of meaning within a plurality of contexts of use. This is precisely what I call a hermeneutic perspective for the philosophy of information.

Lucas D. Introna (1993) has been among the first ones to develop a hermeneutic approach to information.[11] In the field of studies in the information systems,[12] there are numerous definitions of information that have, for him, at least three points in common: (1) information is understood as the result of a process of conversion or transformation. Data must be converted into information; (2) the information has a receiver or a user. The receiver must have experience of what she receives as something meaningful or valuable; (3) the purpose of information is to cause changes or affect the choices of the receiving system. These three points can be summarized in the statement that "the only condition for data to become information is that it must be meaningful to the recipient" (np). The interpretation serves to correct errors in the communication process. A sender usually encodes her message in symbols, words, signs, et cetera, to communicate to a receiver some meaning that she thinks the receiver should know. The receiver must unlock or interpret these symbols, words, or signs, but in this process, she is subjected to numerous sources of noise that undermine her success. Among the possible sources of noise, the author suggests (1) a limited set of symbols compared to an unlimited series (or at least much broader) of meanings (this is the case of every spoken language); (2) a temporal distance between the sending and receiving of the message that can vary from a few seconds to millennia (remember the example of the Rosetta Stone); (3) a changing series of relations between words, symbols, or signs and their meaning (think about the variation of languages over time, such as when you find yourself reading Dante's *Divine Comedy* in thirteenth-century Italian, or moving from one language to another); (4) different reference contexts or contexts used by the sender and the receiver during the communication process (for example, two people who speak the same language but belong to very different cultures, social backgrounds, or generations).

According to Introna, it is not a question of secondary or exceptional cases: in a certain sense, it can be said that any communication is by nature "failed" and therefore needs interpretation and hermeneutics. The difference, as already stated, is between white or dead interpretations and living interpretations. The fact that the origin of an interpretation has been forgotten does not mean that it is no longer an interpretation. It is also important to note that although Introna develops its argument on a model of information and communication 'in the making,' the same ideas would remain valid in the process of criticism and deconstruction of an information and communication process that has already taken place. Within the information systems literature, Myers (1995, 57) has proposed a dialectic hermeneutic approach, in which the researcher does not simply accept the participants' self-understanding but tries to evaluate the totality of understandings in a given situation critically. Finally, it must be said that (good) interpreting does not necessarily mean tracing the message back to the intentions of its author. As Introna himself says (1993, np), it would be a pity if the only possible meaning that could be attributed to a work of art was the one the creator intended to give to it at the time of its creation. Incidentally, works of art are a particular case of information systems in which the meanings can be many, certainly more than they can be in a sentence of an ordinary language, but even in this case, they are not infinite.

The author also insists on the notion of hermeneutic circle, according to which the interpretation, for example of a text, is based on tradition, context, and, more generally, the specific worldview from which one interprets. In the case of two persons in dialogue, one can think of the interpretation of information also as the negotiation between these two worldviews that contend each other, so to speak, the meaning of the message.

In the conclusion to the article, Introna claims that the information has the following characteristics: (1) its understanding is based on lived experience; (2) is the result of a process of appropriation; (3) is contextual; (4) is perspective; (5) is never complete; (6) it cannot be transmitted from one person to another, because only data can be transmitted. This point is particularly interesting because it means to say that in the world there is no information, but only in the minds, preferably the human ones. In a sense, in the integrated circuits of a computer there is no information, no words, no letters, and even no 0 and 1, but just electricity directed in specific ways that leads us to interpret it as information; (7) is an intrinsic part of the communication or the total organizational dialogue and cannot be artificially separated from it. In short, while sharing Floridi's perspective about what is philosophically relevant in the information (indeed, for both of them, semantics—and pragmatics—are the informational issues par excellence), Introna also indirectly shows how the main problem in the STI is precisely the one the philosopher of Oxford immediately rejects: the relationship between message, mind (used here in the vaguest way possible), and context.

Oscar Diamante (2014, 181) has similarly argued that the question of information should be included within the hermeneutic issue of understanding. It is for him, in a sense, a matter of reincluding information into that long chain of knowledge that goes from data to understanding and eventually to wisdom.[13] Sometimes this understanding is merely in the background, as when a habitual user deals with the computer mouse, while at other times it is in the foreground when, for example, we have to confront new technological objects. Think of the hard work of the designer who reshapes objects of everyday use, such as a chair or a graphical user interface. Incidentally, this example is interesting because it makes us understand how hermeneutics is not to be seen as a merely cognitive operation. There is, in fact, material interpretation in design, as well as in many other practical activities in the world.

To return to the topic in question, Diamante (184) observes: "information cannot be information [. . .] if there is no question (however tacit) to which it is an answer. [. . .] If the thermometer reads a room temperature of 33°C, this information appears so within the frame of questioning, say, about the temperature for the day." The information, therefore, depends on the presence of an interested observer, an interest that derives from a specific context, so that the information will bring with it a certain degree of relativity. In this regard, one must insist on the fact that contextuality and incompleteness of information do not lead to radical relativism or, to use the critical formula that I have applied to Floridi, a night in which all cows are interpretational. Two limits, in fact, regulate the interpretive multiplicity: an upper limit, which in the triadic semiotics of Peirce would be the thirdness of the interpretant, and a lower limit, which is the reality itself, for which I am free to call, use, and even perceive a thing in many ways, which however are not infinite. Umberto Eco (2012, 98) recalls a dialogue in Cambridge in 1990 with Richard Rorty, according to whom although it is possible to interpret a screwdriver as a tool for screwing, it is also legitimate to see it and use it as a tool to open a package. Eco bases his objection on Gibson's notion of affordances, that is, the properties that a particular object exhibits, and which make it more suitable for some uses rather than for others. In fact, if the screwdriver can be seen and used with equal legitimacy to screw or to open a package, it is strongly discouraged to use it to clean the ear, so it is much better to use a cotton bud. The same holds true for a painting, a text, and a database: we can say and do many things with them, but their possible uses are not infinite.

* * *

In conclusion, the hermeneutics of information (and, more broadly, digital hermeneutics) is a material hermeneutics for three reasons: (1) because it starts from an analysis which is internal and not extrinsic to the object in question; (2) because it deals with the varieties of contexts of production and reception of meaning; (3) because it is interested in the matter (the techniques and technologies) through which digital traces are transformed into data, and data into information. The first two elements have been introduced and partially articulated by Peter Szondi (who used the expression "material hermeneutics" long before postphenomenologists) in his *Introduction to Literary Hermeneutics* (1995) in the specific and indeed too narrow context of philology and texts interpretation. The third one can be considered as a critique of Floridi's diaphoric definition of data. Hermeneutics of data, indeed, concerns, among other things, the material conditions of data and database structuration—see the conclusion of this part of the book for a further discussion of this point. For the moment, we can say that for digital hermeneutics, information is never a starting point, but always a renegotiable point of arrival.

In an interview given to me in 2016, Floridi discusses the notion of "onlife" starting from the example of a person walking in the city using the GPS, listening to music online and writing to a friend via Whatsapp at the same time: "does it make sense to ask if I am online or offline? You must be at least fifty years old to ask yourself a similar question. If you are younger, it does not even occur to you" (Floridi and Romele 2016, 170). The fact is that, as he says soon after, "we spend more and more time in this sort of delta of the river where the fresh waters meet the salty ones. Obviously, there are the sea and the river. [. . .] But what we must consider is that this our conversation [via Skype] is happening in a space that is neither online nor offline, but it is a bit both." And so, Floridi concludes, "it is better to stop placing this delta either in the river or the sea and to understand what this life a bit paludal is and what is this fauna that is emerging in an increasingly large environment." Now, a hermeneutic perspective does not want to return to the alternative between the real and the virtual; neither does it want to forget such a distinction. Its aim is instead to focus on their articulations, which means on their similarities but also differences.

Notes

1. In this context, I am putting aside most of the history of information and its relation to the development of computers and digital machines.
2. In 1962 in Royaumont, Simondon underlines in dialogue with Wiener how in his approach, information as communication has nothing to do with the emission or with the transmission but with the reception: "information is not a thing, but the operation of something that arrives in a system and produces a transformation. Information cannot be defined outside this act of transforming incidence and the reception operation" (Simondon 2010, 159).
3. In Floridi (2010, 20–22), the Oxford philosopher talks about the General Definition of Information, GDI. As for the syntax, he writes that "[s]yntax here must be understood broadly, not just linguistically, as what determines the form, construction, composition, or structuring of something." He also writes that "[h]ow data can come to have an assigned meaning and function is a semiotic system like a natural language is one of the hardest questions in semantics, known as the *symbol grounding problem*. Luckily, it can be disregarded here." But this problem can only be neglected within a theory in which meaning and truth are already "encapsulated" in the definition itself. They become, on the contrary, fundamental in the hermeneutic perspective I am proposing here.
4. In Floridi (2015, np), the author also speaks of "Typological Neutrality," TN), already introduced, but without using this term, in previous publications, where he distinguishes between five types of data: primary, secondary, metadata, operational data, and derivative data.
5. In Floridi (2010, 32), he presents environmental information through the example of the trees' growth rings. Another example he uses is taken from the crime television series *CSI*, in which it is often a matter of bullet trajectories, blood spray patterns, organ damages, fingerprints, and other similar clues that, I would say, become evidence only thanks to the interpretation of smart (all too smart to be real?) detectives.

6. I am referring to Ricoeur's comments to Derrida's "Whyte Mythology" in the eighth study of *The Rule of Metaphor* (2004).
7. In Floridi (2013, 31), one can find the following definition: "A LoA is (usually) a finite but non-empty set of observables, which are expected to be the building blocks in a theory characterized by their very choice." He also introduces the idea of an "interface" or "gradient of abstractions," which "consists of a collection of LoAs and is used in analyzing a system from varying points of view or at varying LoAs."
8. https://ec.europa.eu/digital-single-market/sites/digital-agenda/files/Manifesto.pdf. Accessed June 10, 2019.
9. The author specifically refers to communism, transhumanism, and rational Christianism—"to pray to God, is to program the system; the justice of God is the immanence of the code; and the consolation of God is the triumph over Death" (Alizart 2017, 181).
10. J. Searle, "What Your Computer Can't Know." *New York Review of Books*, October 9, 2014. www.nybooks.com/articles/2014/10/09/what-your-computer-cant-know/. Accessed June 10, 2019. According to Searle, the problem with Floridi's theory of information is that "there is nothing intrinsic to the physics that contains information. The distinction between the observer-independent sense of information, in which it is psychologically real, and the observer-relative sense, in which it has no psychological reality at all, effectively undermines Floridi's concept that we are all living in the infosphere"; "Almost all of the information in the infosphere is observer-relative. Conscious humans and animals have intrinsic information, but there is no intrinsic information in maps, computers, books, or DNA, not to mention mountains, molecules, and tree stumps."
11. In this sense, the works of Capurro are also forerunners, including Capurro (1992, np), in which information is presented as a "hermeneutic-rhetorical discipline, [which] studies the *con-textual* pragmatic dimensions within which knowledge is shared *positively* as information and *negatively* as misinformation particularly through technical forms of communication."
12. There is a particular interest in hermeneutics in the studies on information systems. See above all Webb and Pollard (2006) for a good summary of existing approaches.
13. According to Rowley (2007) the first occurrence of the "wisdom hierarchy" (data, information, knowledge, wisdom) can be found in T. S. Eliot's poem *The Rock*: "Where is the wisdom we have lost in knowledge? / Where is the knowledge we have lost in information?" For Bellinger, Castro, and Mills (2004), understanding is not a separate level in this chain, as suggested by other authors, but supports the transition from each stage to the next.

References

Adam, A. 2008. "Ethics for Things." *Ethics and Information Technology* 10(2–3): 149–154.

Adriaans, P.W. 2012. "Information." *Stanford Encyclopedia of Philosophy*. https://plato.stanford.edu/entries/information/. Accessed June 10, 2019.

———. 2010. "A Critical Analysis of Floridi's Theory of Semantic Information." *Knowledge, Technology & Policy* 23(1–2): 41–56.

Adriaans, P.W., and Vitanyi, P.M.B. 2009. "Approximation of the Two-Part MDL Code." *IEEE Transactions on Information Theory* 55(1): 444–457.

Alizart, M. 2017. *Informatique céleste*. Paris: P.U.F.

Bellinger, G., Castro, D., and Mills, A. 2004. "Data, Information, Knowledge, and Wisdom." www.systems-thinking.org/dikw/dikw.htm. Accessed June 3, 2019.

Brey, P. 2008. "Do We Have Moral Duties Towards Information Objects?" *Ethics and Information Technology* 10(2–3): 109–114.

Capurro, R. 2009. "Past, Present and Future of the Concept of Information." *tripleC (Cognition, Communication, Co-operation)* 7(2): 125–141.

———. 1992. "What Is Information Science For? A Philosophical Reflection." www.capurro.de/tampere91.htm. Accessed June 3, 2019.

De Mul, J. 1999. "The Informatization of the Worldview." *Information, Communication & Society* 2(1): 69–94.

Diamante, O.R. 2014. "The Hermeneutics of Information in the Context of Information Technology." *Kritike* 8(1): 168–189.

Eco, U. 2012. "Di un realismo negativo." In M. De Caro, and M. Ferraris (eds.). *Bentornata realtà. Il nuovo realismo in discussione.* Turin: Einaudi, 91–112.

Ess, C. 2009. "Floridi's Philosophy of Information and Information Ethics: Current Perspectives, Future Directions." *The Information Society* 25(3): 159–168.

Floridi, L. 2015. "Semantic Conceptions of Information." *Stanford Encyclopedia of Philosophy.* https://plato.stanford.edu/entries/information-semantic/. Accessed June 10, 2019.

———. 2014. *The Fourth Revolution: How the Infosphere is Reshaping Human Reality.* Oxford: Oxford University Press.

———. 2013. *The Ethics of Information.* Oxford: Oxford University Press.

———. 2011. *The Philosophy of Information.* Oxford: Oxford University Press.

———. 2010. *Information: A Very Short Introduction.* Oxford: Oxford University Press.

———. 2008a. "The Method of Levels of Abstraction." *Minds and Machines* 18(3): 303–329.

———. 2008b. "Information Ethics: A Reappraisal." *Ethics and Information Technology* 10(2–3): 189–204.

———. 2005. "Is Semantic Information Meaningful Data?" *Philosophy and Phenomenological Research* 70(2): 351–370.

Floridi, L., and Romele, A. 2016. "Filosofia dell'informazione e tracce digitali: Alberto Romele intervista Luciano Floridi." *Azimuth* 4(7): 169–175.

Grodinsky, F., Miller, K., and Wolf, M.J. 2008. "The Ethics of Designing Artificial Agents." *Ethics and Information Technology* 10(2–3): 115–121.

Iliadis, A. 2013. "Informational Ontology: The Meaning of Gilbert Simondon's Concept of Individuation." *communication +1*, 2, article 5.

Introna, L. D. 1993. "Information: A Hermeneutic Perspective." *Proceedings First European Conference of Information Systems*, Henley on Thames (UK).

Myers, M.D. 1995. "Dialectical Hermeneutics: A Theoretical Framework for the Implementation of Information Systems," *Information Systems Journal* 5(1): 51–70.

Peters, J.D. 1988. "Information: Notes Toward a Critical History." *Journal of Communication Inquiry* 12(2): 9–23.

Ricoeur, Paul. 2004. *The Rule of Metaphor.* New York and London: Routledge.

Rowley, J. 2007. "The Wisdom Hierarchy: Representations of the DIKW Hierarchy." *Journal of Information Science* 33(2): 163–180.

Shannon, C. 1948. "The Mathematical Theory of Information." *Bell System Technical Journal* 27(3): 623–666.

Shannon, C., and Weaver, W. 1964. *The Mathematical Theory of Communication*. Urbana: The University of Illinois University Press.

Simondon, G. 2010. *Communication et information: Cours et conférences*. Chatou: Les éditions de la transparence.

———. 2005. *L'individuation à la lumière des notions de forme et d'information*. Grenoble: Millon.

Szondi, P. 1995. *Introduction to Literary Hermeneutics*. Cambridge: Cambridge University Press.

Webb, P., and Pollard, C. 2006. "Demystifying a Hermeneutic Approach to IS Research." *Australasian Journal of Information Systems* 13(2): 31–48.

Wiener, N. 1989. *The Human Use of Human Beings: Cybernetics and Society*. London: Free Association Books.

———. 1965. *Cybernetics. Or Control and Communication in the Animal and the Machine*. Cambridge, MA: The MIT Press.

2 The Real Invaded the Virtual

The idea of the virtual as a place apart from reality dominated the literature of the 1980s and 1990s. According to danah boyd (2001, 3–4):

> In the late 1980s and early 1990s, many academics imagined that virtual environments would offer a utopian world where sex, race, class, gender, age, and sexual orientation ceased to be relevant. [. . .] As digital pioneers, Donna Haraway, Sandy Stone, and Sherry Turkle imagined the possibility of life online as a way to transcend physical identity and marked bodies. Cyberspace became a site, or series of sites, in which identity might be deliberately and consciously performed (*à la* Judith Butler).

In her *Life on the Screen*, for example, Sherry Turkle presented several cases where simulation games and MUDs (Multi-User Dungeons) gave users the opportunity to experiment different roles and situations, usually with positive effects IRL: "The anonymity of most MUDs [. . .] provides ample room for individuals to explore parts of themselves. [. . .] MUDs imply difference, multiplicity, heterogeneity, and fragmentation. [. . .] MUDs thus become objects-to-think-with for thinking about postmodern selves" (Turkle 1995, 185).

The birth and development of the social Web during the 2000s has increased the realism and reliability of the information provided, while making concealment more complex. Donath and boyd (2004, 72) presented social networking sites as

> online environments in which people create a self-descriptive profile and then make links to other people they know on the site, creating a network of personal connections. Participants in social networking sites are usually identified by their real names and often include photographs; their network of connections is displayed as an integral piece of their self-presentation.

For them, "the public display of connections found on networking sites should ensure honest self-presentation because one's connections are linked to one's profile; they have both seen it and, implicitly, sanctioned it" (Donath and boyd 2004, 73–74). As French sociologist Dominique Cardon (2008, 20) has argued, personal engagement on social networking sites ended up with a general loss of privacy. In this regard, Zhao, Grasmuck, and Martin (2008) have spoken of a shift from anonymity to "nonymity." According to them, "the online world is not entirely anonymous. Family members, neighbors, colleagues, and other offline acquaintances also communicate with each other on the Internet" (1818). More recently, danah boyd (2014, 20) has stated that "[t]he social media tools that teens use are direct descendants of the hangouts and other public places in which teens have been congregating for decades." In other words, social media are just the continuation of socialization IRL by other means, mainly because often today many North American and European teens enjoy less geographic freedom, less free time, and more rules than when we were their age. The virtual has ended because all that happens in the 'virtual' has a real value and actual consequences on our existences.

According to Richard Rogers (2013, 23), the "end of the virtual" can be traced back to a specific event, the case of LICRA (International League Against Racism and Anti-Semitism) and UEJF (Union of Jewish Students of France) against Yahoo! in 2000. These two organizations found out that it was possible to buy Nazi objects at auctions from France through yahoo.com, in violation of the French penal code. They decided to file a complaint against Yahoo! with the Tribunal de Grande Instance of Paris. On November 21, 2000, the court sentenced Yahoo! "to take all measures likely to discourage and render impossible any consultation on yahoo.com of the auction services of Nazi objects and any other site or service which constitutes an apologia for Nazism or a questioning of Nazi crimes."[1]

The technical repercussion of this verdict was the development of technology that geographically locates an IP address to direct content nationally. Today, by accessing google.com on a web browser in France, one is automatically sent to google.fr. Similarly, the content of platforms like Youtube, Spotify, and Netflix are different for each country. This territorialization of the Web (and it would be interesting to know what Carl Schmitt would say of a 'sea of information' re-conquered by the land) is today evident, among other things, in European policies on the right to be forgotten which determines variations, on a geographical basis, of online information access. The same holds true for the more recent decision in the United States to renounce to the principle of the Net neutrality.

Most people have a decidedly negative vision of the end of the virtual, on a scale ranging from lukewarm nostalgia to the announcement of an imminent or already active technological apocalypse. According to Boris Beaude (2014), the end of the Internet as we know it can be guessed from various points:

1. Firstly, we must acknowledge the abolition of space and the re-emergence of territory. The author understands space à la Leibniz, that is, something that concerns the relations among entities or the order of things rather than their location. For him, the Internet is a 'real' space precisely in that it can reconfigure the relations among entities. His intention (Beaude 2012, 14) is to defend the Internet's space from the "dominant territorial paradigm," namely, from the "materialism that confuses the real and the material, space and the territory." The battle is not only theoretical; it is also political and ideological, against those territorial forces, such as national states, international institutions, and private enterprises, that are trying and succeeding into territorializing (or 'grounding') the space of the Internet.

2. Secondly, we are witnessing the end of the freedom of speech and the transformation of the Internet into a global Panopticon. After Edward Snowden, who revealed the details of several US and British mass surveillance programs like PRISM, we are *certain* that we *can* be spied at any moment through a massive collection of Internet data. According to Foucault (1995, 202), the Panopticon operates mainly by dissociating the couple see–being seen: "in the peripheric ring, one is totally seen, without ever seeing; in the central tower, one sees everything without ever being seen." In other words, its effectiveness lies not in the actual exercise of power, but in knowing that its power can be exercised at every moment. In this sense, the revelations of Snowden are not the negation but rather the full realization of the global panoptic project, if there is such a thing.

3. Thirdly, we have gone from the Internet of collective intelligence and shared skills to new forms of exploitation and "digital labor" (Scholz 2012; Fuchs 2014): under-qualified work (Amazon Mechanical Turk), often unaware (reCaptcha), and sometimes with the aggravation of mixing with the pleasure of sociability (Facebook's Likes). Moreover, it must also be added that the illusion of independence and self-determination leads people to be exploited by platforms like Uber and Foodora.

4. Fourthly, the Internet is no longer the kingdom of gratuity. If platforms for sharing online music such as Napster symbolized it, today's successful paying services like Spotify perfectly illustrate the end of digital gifting online. More and more people are inclined to pay for services and products they considered free until a few years ago (Romele and Severo 2016). In 2017, for instance, Google launched a new version of Google Contributor, a service first introduced in 2014, which allows users to pay for not seeing ads on their favorite sites.

5. Fifthly, we know that the Internet was born as a military project, but that almost immediately, perhaps thanks to its "intrinsic politics" (Winner 1980), that is to say, an inherently anti-hierarchical network

structure, was adopted by the American Left and academic researchers as a model and medium for sharing and disseminating knowledge. It is on this decentralized infrastructure that new centralities have been built at the scale of the Web. In principle, one may visit any site and become a protagonist on this stage; in reality, we always visit the same sites, and most of us remain passive and silent spectators. The Web practices imply a hierarchical principle of recognition of authority as Larry Page and Sergey Brin well understood when they invented the Google PageRank, which is based on hyperlinks and not, such as old search engines like AltaVista, on indexing. And as those who conduct academic research know (PageRank is actually based on the scientometric model), authority creates even more authority, and the "last will remain last, if the first are unreachable."[2]

6. Finally, the Internet is no longer a place of separation and protection, especially of privacy, but an exposure, and probably even intensification, of the vulnerability of individuals. Our lives are immediately put on display to the performative gaze (Foucault emphasizes precisely this double nature, 'epistemological' and 'ontological', of the Panopticon) not only of the others but also of the 'Others,' that is, the sociotechnical systems.

There are several reasons to agree with most of these points, and many of them will be further discussed in the following pages. However, in this chapter, I would like to focus on another, mainly epistemological consequence of this "end of the virtual." My argument will be developed in three sections. In the first section (2.1), I will account for what has been considered the main advantage of the fact that the real has invaded the virtual. I am referring to digital representations that have started being considered as reliable and exhaustive representations of the social reality. In the second section (2.2), I am going to propose an intermezzo by discussing the few but somewhat influential ideas of Bruno Latour on the digital. He sees a sort of strong homology between social reality (at least, his version of it) and the digital. Moreover, Latour finds in digital methods for social research adequate means to scientifically approach social reality as he understands it. In the third section (2.3), more cautious approaches in the field of digital sociology are going to be considered, which are closer to a hermeneutic approach to the digital.

According to Marres (2017, 12–17) "digital sociology" may refer to: (1) the study of the presence of digital technologies in society; (2) the transformation of society through digital technologies; (3) the study of society through digital methods; (4) the transformation of society through its study with digital methods. Here I am using the term mostly according to its third meaning. I will also insist on the specific social ontology inspired by or at least sympathetic to the use of digital methods for social research.

2.1 Digital Sociology, Part One

The two movements according to which "the virtual invaded the real" and "the real invaded the virtual" are not sharply delineated as I have suggested up to now. The idea that the real has invaded the virtual refers, firstly, to the fact that what happens online is as effective in our lives as what happens offline. For instance, a Like on Facebook or a Mention on Twitter are no less relevant than offline and face-to-face manifestations of mutual recognition. Secondly it means that an increasing part of our actions has what might be called their digital shadow or ghost—remember Derrida's "hauntology." In both cases, this happens because the digital has invaded the real. Thanks to the digital, we can make 'real' things, such as purchasing a plane ticket or having a conversation with friends. Furthermore, the diffusion of mobile and wearable technologies, along with the presence of connected objects and smart environments ensure that our actions are increasingly being digitally tracked. In the course of this book, I will repeatedly affirm that the 'new' of new technologies does not lie in information and communication, but rather in registration and recording.

According to Richard Rogers, the end of the virtual so understood is not entirely bad. Indeed, there is at least a substantial advantage, namely, that one can resort to digital traces to study social reality, without going back to "the things themselves" anymore. If opportunely treated, digital traces become reliable representations of the reality. In the words of Rogers (2013, 21)

> I would like to help define a new era in internet research, one that no longer concerns itself with the divide between the real and the virtual. [. . .] The Internet is employed as a site of research for far more than *just* online culture. The issue no longer is how much of society and culture is online, but rather how to diagnose cultural change and societal conditions by means of the Internet.

The author distinguishes between two approaches in the field, digital methods and computational sociology.[3] Such a distinction is not without analogies with the one between the *bricoleur* and the engineer I am going to discuss in the second part of the book. The latter is exemplified by the Google Flu Trends project. Launched in 2008 but now interrupted, the Google Flu Trends was a web service operated by Google which provided estimates of influenza activity for over twenty-five countries by aggregating Google search queries. As David Lazer and Ryan Kennedy explained in an article published in *Wired* in 2015, the essential idea behind the Google Flu Trends was that when people are sick with the flu, many of them search for flu-related information on Google, providing almost instant signals of overall flu prevalence. This methodology was faster than classical methods: "search data, if properly tuned to the flu tracking information from the Centers for Disease Control and Prevention, could produce accurate estimates of flu prevalence two weeks earlier than the CDC's data."[4] In Ginsberg et al. (2009), the web service was for the first time presented as follow:

> Because the relative frequency of certain queries is highly correlated with the percentage of physician visits in which a patient presents with influenza-like symptoms, we can accurately estimate the current level of weekly influenza activity in each region of the United States, with a reporting lag of about one day. This approach may make it possible to utilize search queries to detect influenza epidemics in areas with a large population of web search users.

It was an authentic work of Big Data and big social computing since it was performed by processing hundreds of billions of individual searches from five years of Google web search logs. The amount of computation had been considerable as well: "By aggregating historical logs of online web search queries submitted between 2003 and 2008, we computed time series of weekly counts for 50 million of the most common search queries in the United States." Especially considering that the researchers fitted 450 million of different models to test each of the candidate queries: "the amount of computation required could have been reduced by making assumptions. [. . .] However, we were concerned that aggressive filtering might accidentally eliminate valuable data" (4).

The other approach, the one Rogers privileges and practices, is exemplified by a work of infographics and data visualization published in 2009 in the *New York Times* from the queries on the website Allrecipes.org one day before the Thanksgiving. These data were captured and plotted on a geographical map, showing the locations of recipe queries across the United States. The exciting aspect of this work is that it did not merely deal with the most consulted recipes of the month or the year, but that the recipe query maps also added a new social dimension: it displayed where people seemed to like which food. Even in this case, numbers were not small: "by 9 p.m. Wednesday [the day before Thanksgiving], 785,000 people had looked up turkey recipes at Allrecipes.com. For most of the day, the site was handling one million page views an hour," and "[b]y noon, 100,000 people had searched for mashed potato recipes."[5] But the amount of work and the methodologies implied are not comparable in terms of complexity, automation, and technicity with those that were necessary in the case of the Google Flu Trends. It is a bit like the story of David against Goliath: on the one hand, a big company funding big research teams and brutal computation; on the other hand, a small research group such as Rogers's Digital Methods Initiative whose interests are geared more toward media studies, but with a peculiar taste for "beautiful data" (Halpern 2014). "Follow the medium" is the motto of the digital methods, which means that the social researcher must be content with the tools and the objects the medium itself offers.

The difference between digital methods and computational sociology, thus, is a matter of size and skills, but also of attitude and expectations. According to Lazer et al. (2009, 721), the capacity to collect and analyze massive amounts of data, which has already transformed disciplines such as biology and physics, is finally also transforming sociology: "computational social science is occurring—in internet companies such as Google and Yahoo, and in government agencies such as the U.S. National Security Agency." Compared to the classic social research and based on "one-time, self-reported data," new technologies "such as video surveillance, e-mail, and 'smart' name badges, offer a moment-by-moment picture of interactions over extended periods" (722). Researches in this field will bring, according to the authors, several improvements in social relations and structures. E-mail data could reveal dynamics in group interaction ("What interaction patterns predict highly productive groups and individuals?"); face-to-face group interactions could be assessed over time with "sociometers"; "social network websites offer a unique opportunity to understand the impact of a person's position in the network on everything from their tastes to their moods and their health," and so on (722). Despite several obstacles to realizing their epistemological dream (contrary to Quarks and cells, people alas tend to mind when "we reveal their secrets"), these computational scientists seem slightly optimistic. They are neither bothered in citing the NSA among the good practices and opportunities in the field—in their defense, it must be said that the existence of the PRISM surveillance program, started in 2007, was not yet leaked in 2009. Nor are they concerned about the several economic and technological issues, since computational social sciences imply, among other things, a strong centralization even for those 'soft' sciences that could be previously performed adequately also by small research units.

But this is not my primary concern here. Instead, I want to reflect on the epistemological issue concerning the alleged homology between social reality and its digital representations. For example, in his acclaimed book *Social Physics* (2014), Alex Pentland—who is one of the leading representatives of the computational social sciences—exalts cooperation and imitation among individuals as a winning model over competition and individualism: "[t]he basic concept is that by pooling ideas, we can get an average 'wisdom of the crowd' judgment that will be better than the individual judgements" (28). Incentives and different forms of nudging can be used to improve mutual engagement instead of isolation. Certainly, nobody would disagree that cooperation is better than waging war. But the price to pay for such a praiseful perspective is very high, since it consists of the total predictability—and control, in the sense of Wiener—of human societies. Such predictability can be achieved thanks to massive data collection and processing of which Pentland offers several examples in his book. Thus, he speaks of the need for a "New Data Deal," in which "individual rights to personal data must be balanced with the needs of corporations and governments to use certain data—account activity, billing information, and so on—to run their day-to-day operations" (181). For him, "the potential rewards of a data-rich society operating on the principles of social physics are worth the effort and the risk" (216).

I believe there are two sorts of risk in such a statement. The first one, which is also the most obvious, is related to surveillance and the general loss of privacy. We all know about the several ongoing projects of "social physics" in China, such as massive face recognition or the Social Credit System. The second one is the society orienting itself toward a model—the one made of collaboration and collectivization of decisions—that makes it digitally understandable and predictable. In other words, digital technologies and data are used for both 'reading' and 'shaping' social reality.

Among the several tools experimented by Pentland and presented in his book, there is, for instance, the Meeting Mediator. This tool has two main components: a sociometric badge to capture turn-taking behavior and a mobile telephone to visualize the group's interactions. When everyone contributes equally to the conversation, a ball moves at the center of the telephone's screen. When one person dominates it, the same ball moves toward the person who is speaking too much (108). The Meeting Mediator can be considered as an instrument for improving governance in the sense of participation of all individuals involved in the debate and hence to the decision making. But it can also be seen as a tool for new forms of governmentality, in which surveillance and control are exercised by institutions/organizations on social groups and individuals who are inclined to shape their behaviors on the basis of allegedly neutral quantifications. As I will extensively argue in the second part of this book, I do believe that human beings are to a large extent predetermined in their own decisions and actions, as I think that the recent development of digital technologies has profoundly changed the representation human beings have of themselves. Under certain conditions, I am also sure that such tools might be particularly helpful for practicing a sort of digital hermeneutics of the self. I cannot prevent myself, however, from believing that there is always an inexhaustible gap between the self and its digital representations, as there is a gap between auto- and hetero-determinations. This will be the main topic in the finale. Here, I want to point out more modestly to the fact that digital sociology has not been able to realize up to now its promises. The recent history of digital sociology remains for the moment the history of a disillusion.

However, accusing the computational social sciences of minimizing the risks and having several technological and epistemological limits is rather easy. For this reason, once again, I have decided to take a more challenging route, which consists of Bruno Latour and his few but important reflections on the digital. Despite their differences, I believe that Latour's approach and computational social sciences have been for a while both victims of the same illusion of transparency.

2.2 Intermezzo: Bruno Latour

Bruno Latour is for sure not a digital sociologist. And yet, his studies and his few considerations on the digital had a significant impact, both theoretical and practical, on several digital fields. My thesis is that his perspective on the digital, to which he resorts to endorse his view of social reality as an actor-network, contradicts some of his philosophical and sociological ideas. To be more radical, I would also say that his insights on the digital, in which he finds support for the actor-network theory, have the opposite effect of demonstrating its partial inconsistency.

One might think that Latour's proverbial attention to matter and technologies would have led him to an attentive analysis of the continuous discrepancies between the 'real' and the 'virtual,' between the social reality and its digital representations. One might think that he would have been interested in the materiality of new media, such as cables, data farms, computers, and slow connections as if to mock the literature on the spaceless space or the fourth revolution. I expected ethnographic reconstruction of the complex and dense network of actants between economy, politics, nature, and technology that has led to the conversion of many nuclear war bunkers in Switzerland into data centers where millionaires from the entire world hide and protect their bitcoins. Nothing more 'real' and 'heavy' (the entrance doors of these bunkers can weigh up to ninety kilos) for what is considered the virtual and volatile currency par excellence. I imagined Latour sending one of his Ph.D. students to collect interviews from those real and concretely exploited persons hiding behind the Amazon Mechanical Turk. Another student of his would have explored the politics of submarine cables around the world, another one would have observed people in the Paris subway struggling with the lack of signal, trying to call, send messages, and read Internet pages, and so on.

Such has always been the strategy of the French sociologist and anthropologist: to show the matter of spirit and the spirit of the matter, the culture in nature and the nature in culture. In *Aramis or the Love for Technology*, his aim had been to "offer to the humanists a detailed analysis of a technology sufficiently magnificent and spiritual to convince them that the machines by which they are surrounded are cultural objects worthy of their attention and respect" (Latour 1996, VIII). Here, in his meditations on the digital, the aim might have been to show that the 'spirit' of the virtual is grounded in the matter and its affordances. In *Aramis*, Latour (170) speaks of a "supple frame of reference," and, in

Einstein's words, of a "reference mollusk." In this context, he might have accounted for the material support from which the digital sign, although formal and "ascetic," cannot free itself. Again, in *Aramis*, the French sociologist and anthropologist wanted to return the Spirit to the Letter, because mechanisms and technologies, accused by humanists of being 'purely' functional and 'strictly' material, actually hide compromises, desires, Spirit, and of course a lot of morality: "They are the scapegoats of a new religion of Silence, as complex and pious as our religion of the Word (*Parole*). What exegesis will have to be invented to provide commentary on the Silence of the machines?" (206–207. Translation modified). Perhaps, for the Internet and the digital, Latour could have undertaken a contrary movement, that is, he could have tried to bring back the whole Spirit of the digital to its matter, which is the electricity of the power stations, the cables that transmit information, the data farms, subscriptions, passwords, especially those that are forgotten, et cetera. Instead of Floridi's joyful and soft infosphere, Latour would have been able to write about the dark and hard rooms of the internet cafés still existing in many working-class arrondissements of Paris.

In fact, this seems to be the direction in which his considerations on the virtual go in the first talk he gave on the subject, in 1998 at Brunel University:

> Whenever you get near computers, whenever you get near this digitality, you get cables, masses of cables. [. . .] [S]ome myopia is necessary to counter-balance the hype around virtuality. [. . .] I tried to get to the Thames on the Virtual Web Site to which I was sent by my friends. [. . .] [I]t never worked. It was slow, it was boring, it was flat, it was extraordinarily limited. [. . .] We [. . .] constantly think of it as a spiritual, finally disembodied machinery and we constantly discount, in all the meanings of the word, the difficulty, the heaviness, the slowness, the number of the modem, the modem that never works, the handshake that never handshakes and so on and so forth.[6]

Latour is referring to Le Deuxième Monde, a virtual world created in 1997 by Canal+ Multimedia and shut down in 2001. He also speaks of a "modernist hype, that is, the desire of being a spirit at last and no longer inside the materiality of Society. Here again I refer to the strange way we have of talking about all this technology as if it has no defect, as if it was not slow, as if it was easy and user friendly."

And yet, in his more recent reflections on the digital, there is less "love for technology" and more interest for "reassembling the social." In short, Latour's interests in the digital are above all related to what the digital does or, even better, what the digital shows of society. My thesis is that the second element tends to conceal the first one. At the beginning of his

Prince of Networks: Bruno Latour and Metaphysics, Graham Harman (2009, 5) writes:

> As often happens with the most significant thinkers, Latour is attacked simultaneously for opposite reasons. For mainstream defenders of science, he is just another soft French relativist who denies the reality of the external world. But for disciples of Bloor and Bourdieu, his commerce with non-humans makes him a sellout to fossilized classical realism.

For sure, I do not share either of these two critiques. And yet, I do believe that Latour is both a materialist and a relativist (in the best sense of the word you can imagine). Latour's metaphysics is twofold; it is a metaphysics of the matter *and* a metaphysics of the networks. While usually this double metaphysics is in elegant balance (a delicate balance to which, incidentally, I largely agree), it seems to me that where Latour speaks of the digital he tends to favor networks on the matter. As if the digital (re) presentations of social reality were transparent and 'innocent,' as if they were not just able to represent things, but also to 'present' them as they actually are—I am referring here to Gadamer's aesthetics, according to which art is more a "presentation" (*Darstellung*), in the sense that it can reveal its very essence, rather than a mere representation (*Vorstellung*) of reality. It is as if Latour's attention to the matter of the spirit applies to everything except to digital technologies and methodologies because they allow social reality to be seen as Latour wants to see it.

For him, the digital has a double function. From an ontological point of view, it is a model and a paradigm for seeing the society as an actor-network. From an epistemological perspective, it offers a new resource to study society 'in action.' Mainstream sociology has represented social reality in 3-D, that is, distinguishing (at least) between two levels, one of the individuals, and one (to be privileged) of society. But such a perspective, Latour believes, is the result of the methodological limits of the discipline, which was traditionally obliged to simplify social reality to study it. Think back to the surveys making general inferences from a reduced amount of data. Such simplification has brought sociology to hypostatize general movements and tendencies, to believe in the existence, for instance, of social classes and intentionalities.

But things have recently changed. In what might be considered his *summa sociologica*, *Reassembling the Social*, Latour (2005, 119) writes that "the more science and technology extend, the more they render social ties *traceable*. A material infrastructure provides every day more proof of a precise follow up of associations, as any look at the World Wide Web turned World Wide Lab shows."[7] Having the instruments that allow for following the actants without simplifications and generalizations, we can finally understand social reality for what it is, namely a *flatland*: "[I]t's if as we had to emulate in social theory the marvelous book *Flatland*, which tries to make us 3-D animals live inside a 2-D world only made up of lines. It might seem odd at first, but we have to become the Flat-Earthers of social theory" (171–172). Interestingly enough, Latour seems to forget that Abbot's intention in *Flatland* is the opposite. In *Flatland*, indeed, the story is told of A Sphere that visits Flatland at the turn of each millennium to introduce a new apostle (in the novel, A Square) to the idea of a third dimension in the hopes of eventually educating the population of Flatland.[8]

On the same page, Latour also writes: "It's as if the maps handed down to us by the tradition had been crumpled into a useless bundle and we have to retrieve them from the wastebasket. Through a series of careful restorations, we have to flatten them out on a table with the back of our hand until they become legible and usable again."[9] For me, it is as if Latour were erasing the difference between the map and the territory: flattening the maps, they become the territory and no longer some of its possible representations. The Web and, more specifically, digital traceability has transformed reality into a global laboratory in which entities and events can be followed step by step.

Latour's attitude is paradoxical. On the one hand, he realizes how the materiality of the infrastructures, including the Internet, makes relationships and networks visible. For instance, in another passage of *Reassembling the Social* (181) he states: "satellites, fiber optic, calculators, data streams, and laboratories [and, in the French version of the book, he also adds the Internet] are the new material equipment that underlines the ties as if a huge red pen was connecting the dots to let everyone see the lines that were barely visible before." But on the other hand, at least one of these infrastructures, namely the Internet, seems deprived at the end of its materiality.

Consider the 2007 short text "Beware, Your Imagination Leaves Digital Traces," published in the *Times Higher Literary Supplement*. Its title could make one think of a warning, taking position against the attempt of controlling and orienting what is "*interior intimo meo*," that is, imagination. A bit like in Adorno and Horkheimer's critique of the cultural industry, which would have reduced our capacities and possibilities of autonomous schematizations. One could also think of Latour insisting on the gap between imagination and its digitalized traces. And yet, what most interests the French sociologist and anthropologist is the fact that "[t]he precise forces that mould our subjectivities and the precise characters that furnish our imagination are all open to inquiries by the social sciences. It is as if the inner workings of private worlds have been pried open because their inputs and outputs have become thoroughly traceable" (Latour 2007, np). Before the digital, social psychologists were forced to use vague terms and concepts like "rumors," "influences," "fads," "fashions," and "context." Today, the ancient divide between the social and the psychological appears in its nature of "artifact of an asymmetry between the traceability of various types of carries":

> the data bank of Amazon.com has simultaneous access to my most subtle preferences as well as to my Visa card. [. . .] Dozens of tools and crawlers can now absorb this vast amount of data and represent it again through maps of various shapes and colours so that a "rumour" or a "fad" becomes almost as precisely described as a "piece of news", "information", or even a "scientific fact."
>
> (np)

In short, it is the old epistemological dream of psychologism in human and social sciences (that, to be clear, was embraced and then partly rejected by Dilthey) which can finally be realized. Digital traces are thus the precise expression, perfect indeed, of human interiority. In the age of digital traceability, there is no distance any more between inner and outer word. It is as if Latour were too attracted by the possibilities offered by these new technologies and methods to really worry about their material limits.

In 2009, Latour became the director of the MédiaLab at SciencesPo Paris, one of the elite universities in France. His ideas could be further developed in interaction with digital social scientists, mainly those close to the digital methods approach, computer scientists, and digital designers. The MédiaLab found itself at the center of a broad network of institutions more or less consciously spreading all around the world the Latourian ideas and practices—such as mapping controversies, with or without digital methods. In a dynamic of mutual influence, digital researchers have found in Latour's thought an inspiration, while Latour himself has seen a sort of confirmation of his version of the actor-network theory in these researches on the digital.

Invited by Manuel Castells in 2010, he delivered a lecture at the Annenberg School for Communication and Journalism in Los Angeles. In this context, he immediately recognizes that his interest in digital networks does not concern the extension of the various sociotechnical systems of information and communication. He rather considers digital networks as "a powerful way of rephrasing basic issues of social theory, epistemology and philosophy" (Latour 2010a, 2). He speaks, for example, of how networks lead to the reversal of the classic relationship between the substance and its attributes. For Aristotle, attributes are a weak form of being, so much so that they are very close to non-being. And I would say that this weakness is above all of our knowledge. As rational and writing animals (that is, having externalized our rationality through a specific technique, namely writing), we have been by nature and by force led to 'decide,' that is, cut off, limit, and give an inductive or deductive order. The notion of substance, which played a fundamental role in

Western metaphysics, is nothing but a limitation and simplification of reality according to our epistemological and technological possibilities.

However, especially in the last century, we have learned to appreciate and, in part explore the complexity of relations, and to prefer the verb 'to have' to the verb 'to be.' Through Heidegger, distributed cognition, information theory, relational ontology, and, indeed, actor-network theory, we have challenged the classic notions of subject and objectivity. For Latour, digital networks are the first visible and 'concrete' confirmation of this perspective. The study of actor-networks through digital methods is, moreover, an excellent opportunity to overcome our cognitive and technological limits. These methods allow indeed to manage, that is, to visualize and explore the complexity of a network of relations among actants that otherwise we would have to simplify.

Once again, Latour emphasizes the visibility, materiality, and tangibility of the digital networks: "what I like most in the new networks is that the expansion of digitality has enormously increased the material dimension of networks; the digital, the less virtual and the material to given activity becomes" (8). But it does so by denying, or at least, neglecting it, immediately afterward: "what it [the network revolution] does is truly amazing: it dissolves entirely the individual versus society conundrum that has kept social theorists and political theorists busy for the last two hundred years" (9). He further states that the difference between "individuals" and "society" is an artifact of the rudimentary way data are accumulated. And yet, my point is that I do not understand why society as an actor-network represented through digital methods cannot be an artifact as well. Perhaps the limited amount of data has no role? Perhaps the algorithms, with their forces of attraction and repulsion, which is a gravitational metaphor, do not lead to preferring certain perspectives compared to others? Perhaps visualizations, with their beautiful shapes and colors, do not evoke a strange fascination even in the most trained researcher in the field? And, more generally, could one not say that the vision of society as a network misleads on the fact that in society there are, beyond all possible complications and exceptions, also some trends, orders, and classes?

Latour rejects the perspective à la Durkheim, in favor of the heterodox, one of his rivals, Gabriel Tarde. The digital (and, more specifically, the network analysis of digital traces) has finally given a reason to this latter. In his *Monadology and Sociology* (2012, 39), originally published in 1893, Tarde wrote:

> Before the telescope [. . .] was the universal dream not of immutable and incorruptible heavens beyond those known to us? And in the realm of the infinitely small [. . .] does one not still dream of the philosopher's stone in a thousand forms? [. . .] But everywhere where

a scientist digs beneath the indistinction which is apparent to us, he discovers an unexpected treasury of distinctions.

For him, the existence itself is "to differ," and the task of the sociologist does not consist into simplifying social reality by finding eternal and always valid rules, but rather to go beyond "the appearance of uniformity," in order to find "*a diversity whose depths and secrets we have not begun to fathom*" (45. This latter is a citation of the British chemist and physicist William Crookes). Tarde's sociology has been often criticized for its lack of scientificity. However, today, digital methods allow us to "zoom" without losing in terms of accuracy.[10] Indeed, in a certain sense, it is precisely by "zooming" through digital methods that the most authentic scientificity and quantification can be achieved: "Contrary to common wisdom, and exactly as predicted by Tarde, *the more you individualize the more you can quantify*—or else we have to find another name than quantification to describe the phenomenon (is quali-quantitative a possible term?)" (Latour 2010a, 11).[11]

One of the clearest and strongest statements of Latour (and one of his collaborators) concerning the digital and its potential for sociology can be found in Venturini and Latour (2010):

> Thanks to digital traceability, researchers no longer need to choose between precision and scope in their observations: it is now possible to follow a multitude of interactions and, simultaneously, to distinguish the specific contribution that each one makes to the construction of social phenomena. Born in an era of scarcity, the social sciences are entering an age of abundance. In the face of the richness of these new [digital] data, nothing justifies keeping old distinctions. Endowed with a quantity of data comparable to the natural sciences, the social sciences can finally correct their lazy eyes and simultaneously maintain the focus and scope of their observations.

The abundance of digital traces and the development of appropriate methods to collect, analyze, and visualize them would allow us to overcome that eternal gap between the natural sciences and the sciences of the spirit. Digital methods are the end of shame and envy for human and social sciences—and it would be interesting to study how today, in the age of the abundance of data, many natural sciences are following the digital paradigm developed for the first time in the domain of social sciences. In a sense and with the necessary distinctions, we are not too far here from Chris Anderson's famous thesis on "the end of theory."[12] Or rather, here, the only theory that is worth maintaining is the theory that renounces to theorize and is content, instead, to follow individuals step by step. Of course, one might wonder how a theory can give up being a theory, and how observational instruments such as digital methods can be not methods, mediations, and representations. In his *We Have Never Been Modern* (1993), Latour denoted with the word "modern" two orders of practices: the "translation," consisting of creating hybrids, and the "purification," which continuously hides these same hybridizations. Are we not facing such a process right here? Are we not, on the one hand, creating hybrid entities of (social) nature and (digital) culture (digital traces and the methods for their analysis as "presentification" of social reality) and, on the other hand, concealing the very process of creation of these entities?

The embarrassment before such double movement emerges in part, to my knowledge, only in a passage of the 2012 article signed by Latour and some of his collaborators "The Whole Is Always Smaller Than Its Parts: A Digital Test of Gabriel Tarde's Monads." Here the authors once again theorize the existence of a strong analogy, a homology, I would say, between the social world and its digital representations: "It is this experience of clicking our way through platforms such as Flickr™, Academia. edu™ or MySpace, of surfing from document to document, encountering people and exploring communities without ever changing level that we wish to use as an occasion to rethink social theory" (Latour et al. 2012, 592). The model is that of the open monads of Tarde, different from the individual and windowless monads of Leibniz's and Spinoza's modes of a single substance. As Tarde himself asks (2012, 26–27) a question he answers positively, can one think to "open monads which would penetrate each other reciprocally, rather than being mutually external?" For Latour and his colleagues, such "notion may be *rendered fully operational* provided one uses the illustration offered by just the type of navigation through digital profiles we have sketched above" (Latour et al. 2012, 598).

But the question here is: If the Web is what reveals social reality for what it is, what position does the Web have in social reality? And what place do digital methods and researchers using these methods have? They can be exterior or interior, but from a Latourian point of view, both perspectives are problematic. If external, it would mean indeed admitting the existence of a second level in social reality. If internal, it would mean recognizing that the materialized vision of actor-network theory suffers from the same limitations of perspective from which the other visions suffer. In short, the concretization of social reality through the Web would be nothing but a perspective on the city from one of its neighborhoods (to use an example of Leibniz) like the others. The digital researcher and the researcher who is inspired by the Web to get an idea, a model of social reality, would no less be in the magma than any other social actor.

Probably aware of this contradiction, Latour and his collaborators find a bizarre solution, that is of a 1.5 level standpoint, higher and thus distanced from social reality, but not so much to hypostatize it. A bit like the 7.5 floor in *Being John Malkovich*:

> To capture what is none the less a real difference with humans (especially highly scientificized and technicized human collectives), let's say that monads are best captured through a 1.5 level standpoint (1.5-LS). By this expression we mean to say that a) [. . .] a series of intellectual and technical instruments exists to foster the overlapping of different individual definitions; [. . .] and that b) this is what explains the impression that there is "more" in collective actions than what exists in individuals taken in the atomistic sense of the word.
>
> (604)

Of course, this hypothesis is not strange in an absolute sense. I would be willing to say that every interpretation of (social) reality is valid within a specific system and for a specified period. It is a thesis that we can find in Peirce's pragmatism as in Kuhn's theory of scientific paradigms. It is an approach which entertains several analogies with the figure of the hermeneutic circle. But this, I believe, is a bizarre choice within the Latourian system. In fact, all his sociology is not only a theory of social reality but also a meta-theory that wants to criticize all other theories about social reality by denying that there is a society as a whole in addition to individuals. For this reason, the idea of a level 1.5 sounds here to be an exception principle that Latour only applies to his theory.

To conclude, I would say that admitting the existence of a level 'one and a half' means acknowledging, albeit only halfway, at least two things: (1) that digital representations have their effects of "magnification and reduction" as well and (2) that perhaps social reality is not such a flat-land. For this reason, as I said at the beginning of this section, Latour's intuitions on the digital are a gateway to a critique of Latour's sociology as such. I will develop further considerations on this point in the finale.

2.3 Digital Sociology, Part Two

The recent history of Big Data and Big Data analysis is well known.[13] According to Steve Lohr,[14] the first occurrence of the term is in an article by Erik Larson published in 1989 for *Harper's* magazine and then republished in the *Washington Post*. Usually, however, the current use of the term is traced back to the presentation of John Mashey, "Big Data and the Next Wave of Infrastress" (1998). In the academic context, perhaps, the first trace of the term is found at the very beginning of Weiss and Indurkhya (1997, XI): "Very large collections of data [. . .] are now being compiled into centralized data warehouses, allowing analysts to make use of powerful methods to examine data more comprehensively. In theory, 'big data' can lead to much stronger conclusions for data-mining applications, but in practice many difficulties arise." In 2003, Diebold published an article entitled "Big Data Dynamics Factor Models for Macroeconomic Measurement and Forecasting," which he claims (Diebold 2012) to be the first occurrence of the term in statistics and econometrics.

But the real popular success of the expression came in 2008, with the article mentioned in the previous section by Chris Anderson in *Wired* and his thesis according to which "correlation is enough." The term was cleared in the academic context at the same time by Bryant, Katz, and Lazowska (2008). After the hype (the "peak of inflated expectations"), the critical and disillusionment phase began. Consider, for example, the "six provocations for Big Data" by boyd and Crawford (2011). Talking about Anderson—and citing David Berry, who refers in turn to Heidegger—they state:

> Anderson's sweeping dismissal of all other theories and disciplines is a tell: it reveals an arrogant undercurrent in many Big Data debates where all other forms of analysis can be sidelined. [. . .] Instead of philosophy—which Kant saw as the rational basis for all institutions— "computationality might then be understood as an ontotheology, creating a new ontological 'epoch' as a new historical constellation of intelligibility."

In short, even before being a new technology, Big Data would have become a new worldview, a new imaginary, and dominant symbolism, with tangible effects in the way in which individuals make and practice science and undertake the knowledge of their milieu. In the first book of *Metaphysics*, Aristotle recalls that it is not the practical skill that makes wiser leaders, but the fact that they possess the theory and know the causes. But who will be the leaders in a world where we know *that* people buy more Pop-Tarts in the event of a hurricane, but we do not understand *how* and *why* this happens (the example is that of Walmart in Mayer-Schönberger and Cukier 2013, 55)? Regarding the epistemological theme, boyd and Crawford recall not only that "not all data are equivalent" ("Just because two people are physically co-present—which may be made visible to cell towers or captured through photographs— does not mean that they know one another") but also that "bigger data are not always better data." For example, "Twitter does not represent 'all people.' [. . .] Neither is the population using Twitter representative of the global population. Nor can we assume that accounts and users are equivalent" (boyd and Crawford 2011, 6). Rob Kitchin (2014, 2) then said that, in the field of the digital, it is the same term "data" that is problematic and that we should rather talk about *capta* (from the Latin *capere*, "to take"), because they are extracted through observations, computations, experiments, and recordings that have nothing immediate in themselves.[15] Recently, a series of critical approaches to (big) data analysis such as "critical data studies" (Iliadis and Russo 2016) has been initiated.

It is noteworthy that even the same early enthusiasts of the Big Data analysis in social sciences have begun to take more cautious positions. In the context of computational sociology, for example, Lazer et al. (2014, 1203) have warned against what they even call a "Big Data hubris" which consists of their saying in the "implicit assumption that Big Data are a substitute for, rather than a supplement to, traditional data collection analysis." Focusing on the failure of Google Flu Trends, they underline for example the "blue team dynamics" and the "red team dynamics." On the one hand, the algorithm producing the data has been modified by the service provider in accordance with their business model. On the other hand, research subjects attempt to manipulate the data generating process to meet their own goals. The authors conclude the article by saying that Big Data should be counterbalanced by a renewed interest in small data, especially those made possible by the spread of the Internet: standard surveys, experiments, and health reporting. More than a "big data revolution," one should then speak of an "all data revolution.[16]

For digital methods, Venturini et al. (2014) have identified three misunderstandings in the use of digital methods and propose a series of precautions in the use of digital traces in this field of studies. The authors speak in particular of (1) the limits of the representativeness of the data, (2) the fact that these same data have been collected for other purposes such as marketing, surveillance, et cetera, and (3) the belief that data treatments are "objective" and "automatic." The truth is that data are naturally "dirty" (they are, in fact, as we will see below, digital traces); the categorizations used are fragile; the algorithms are poorly documented and must be adapted to the data in use; the visualizations are seductive but easily lead to mistakes. In short, we are before a real (auto)critique of the "scientist arrogance" (Venturini, Cardon, and Cointet 2014, 15) of those who believed in the homology between social reality and its digital representations. It is therefore proposed to make the methods linked to the web data interact with more classical methods such as the old questionnaires. They also recall that every data scientist knows well that there is no method that does not incorporate hypotheses and objectives—nothing closer, in short, of what for many other authors and for me is a hermeneutic attitude toward the digital.

It is as if the digital methods have finally taken an authentically Latourian turn, in the sense of attention to the problems and the material and technological limits related to the use of these same digital methods and tools. And perhaps it is the works of Noortje Marres that prove to be the most Latourian in this field, more Latourian than Latour himself whenever Latour talks about the digital.

In Marres and Weltevrede (2013), the authors are interested, for example, in problems related to scraping, an automated or semi-automated technique for extracting data from websites. "Metaphorically speaking, they state, one could say that scraping structures data collection as a 'distillation process,' which involves the culling of formatted data from a relatively opaque, under-defined ocean of available online materials" (317). Scraping is not a native technique of research in social sciences and must, therefore, be re-oriented if one wants to use it for this purpose. According to the two authors, the main problem is that it is impossible to establish a clear boundary between the medium and its object: "Can we simply understand these devices and platforms as part of our 'methodology,' or should we insist that they are part of the 'object' of our analysis? [. . .] Scrapers, that is, force the question of how we establish the difference between *researching the medium* and *researching the social*" (322).

In short, it is a matter of a hermeneutical technology which, by representing the world, interprets it and which, by interpreting it, also has a blind spot, the one Ihde (1990, 87) calls an "enigma position" between technology and the world. Wanting to look at certain things in the world, the researcher who uses digital methods can only direct his attention to the technological object. Marres and Weltevrede propose two solutions to this problem. The first consists of merely accepting the fluidity of the distinction between objects and methods in this type of researches. Yet, they are not content with this sort of postmodern answer. Instead of accepting and extolling this undecidability, they propose to investigate it and, one could say, make the continuous effort to trace the boundaries between medium, methods, and social reality.

What I find particularly interesting about Marres's perspective is that she does not simply propose an *external* critique of digital methods. This, in fact, is the limit of many philosophical considerations on new technologies that reject the digital and digitalization based on the idea that there is something in life that is not essentially digitizable and quantifiable. In this way, they repropose the old distinction between truth and method, between the sciences of spirit and sciences of nature. Instead, Marres decides to conduct an *internal* critique of digital methods, to inhabit them without being content with accepting their contradictions and limits. In a certain sense, Ricoeur's motto "to explain more is to understand better" seems to resonate in this approach—with the non-secondary difference that in this case, the motto is entirely reversible.

In Marres (2015), this attitude, which implies a kind of faithful unfaithfulness to digital methods, is evident. The author distinguishes two approaches to the problem of digital bias in controversy analysis, one "precautionist" and the other "affirmative." For the former, we need to integrate the use of online data with other offline data. In short, we need to get out of the digital, or at least articulate the digital and the real to reduce the noise. For others, including Marres in fact, it is preferable to embrace this difficulty: "The affirmative approach to digital bias acknowledges and exploits the ambiguity of digital devices, arguing that we can rely on them as empirical means for detecting controversy analysis" (665). Both perspectives recognize that the digital and its methods are not neutral, but only the affirmative one fully recognizes the reconfigurative power of digital representation on the object of analysis—in this case, the controversy, or, another example cited, scientometrics. Being affirmative does not mean being naïve, and therefore we should not be content to say *that* the digital affects the controversy in our society, but also explain *how* this happens. This is what Marres calls "issue mapping": "The investigation of how digital settings influence the public articulation of contested affairs must then become part of our empirical inquiry. Digital controversy analysts should ask not just substantive questions but also formal ones like how is doing issues through data leaks different from doing issues with press releases?" (671). Rather than giving up and putting the digital in brackets because of its limitations, it is a question of embracing it critically, bringing the problem of digital bias at the center of its analysis.

This same theme is developed in Marres and Gerlitz (2015) as well— the ideas of this article are proposed also in Marres (2017), and it is to this latter text that I am referring here. On the one hand, Marres emphasizes the methodological affinities between new and old methods in social research, but, on the other hand, she also observes that there are several divergences between the aims and objectives of digital analytics and sociology: "digital analytics resonate with our own research interests [the ones of the sociologists] but they also invoke alien methodological traditions." And yet, she continues, "rather than seeking to resolve this methodological ambivalence, we may also affirm it, and take it as an invitation to test the methodological capacities of digital devices" (107). The example used is that of Steamgraph, a digital tool which is similar in function to the co-word analysis developed by Michel Callon and his colleagues in the 1980s but which also has important differences. For instance, Steamgraph is focused on "what is happening right now" (liveness), while co-word analysis is rather focused on the emergence of "happening" problems (liveliness). Once again, for Marres it is a matter of finding a third way between the radical criticism and exclusion of the digital and its methods, and their immediate acceptance. For her, one must accept a certain degree of "maladjustment" between methods and medium, but just for implementing "well-chosen adjustments" to "enable maladjustment at the interface between social method and digital settings" (112). I reiterate, the merit of this perspective is to proceed toward an *internal* criticism of the digital. And it is precisely this internal criticism, that is, such attention to both the similarities and the differences, which is profoundly Latourian to me and, I would also say, hermeneutic. Instead of creating a clear-cut separation between two fields of study and being (the human and the digital), this approach seems instead to invoke their articulation.

* * *

The philosopher of technology Gilbert Simondon spoke of "transduction." For him, this notion is closely linked to those of analogy and paradigm. The first is the general logical form of the transductive operation, while the second is the construction or use of a model as a thought process. In *L'individuation psychique et collective* (Psychic and collective individuation) he extended the study of crystals to other fields of knowledge and individuation by transduction. Science, he says, does not usually trust this reasoning because it confuses it with the search for extrinsic similarities and metaphors. But for Simondon this means "to confuse the true analogical reasoning with the method that consists in inferring identity starting from the properties of two beings having in common a character" (Simondon 2005, 108). Instead, the transductive process certainly consists in establishing "the identity of relationships," but these "are not based on similarities, but on differences, and their purpose is to explain them: they tend towards logical differentiation and in no way towards identification" (ibid.).

On the one hand, therefore, the transduction process deconstructs the differences (but also the hierarchies) between different domains of knowledge, being a critique of the classical procedures of induction and deduction—or, to put it another way, being the matrix of a "networking" rather than a "listing" way of reasoning. On the other hand, it also goes a step further, establishing differences even in similarities. Precisely this, one might say, is the attitude of Marres, who embraces digital methods without forgetting their differences—with the methods of classical sociology but also with social reality tout court.

In Simondon, the notion of transduction is not only epistemological but also ontological. The French philosopher has developed a real relational ontology. As he writes, "[a] relation must be understood as a relation in being, a relation of being, a manner of being and not a simple relationship between two terms that we could adequately know through concepts because they would have an actual separate existence" (32). Yet, despite these relations and all the similarities, he does not stop insisting on the differences between the different modes of existence. It is as if Simondon's philosophy was the result of the combination of the theory of relativity, which destroys "the immutable substantiality of the mass" and the quantum theory that in some way re-creates, at another level, "a principle of individuation and stability in discernible beings that the theory of relativity would lose" (131). I wonder if on an ontological level the notion of transduction cannot serve as a corrective to the flatness that characterizes the actor-network theory of Latour.[17]

And more generally, I would tend to say that hermeneutics, with its attention to distanciation even in the articulation, can be not only an epistemological corrective to certain illusions of homology but also an ontological corrective. Here hermeneutics meets postphenomenology (a material hermeneutics, in fact). Both the actor-network theory and postphenomenology try to overcome the classic distinction between subject and object. For postphenomenology, indeed, technologies are not just mediators of the human relation with the world but by mediating they also contribute in the co-constitution of both the human and the world. Yet, while actor-network theory is attentive above all to the plurality of relations, postphenomenology, which usually considers only one relation at a time, addresses somewhat the different types of relations and the different types of actors involved in these relations. In other words, while actor-network theory focuses on the analysis 'in-width,' postphenomenology is rather oriented toward analysis 'in-depth.' According to Verbeek (2005, 165),

> While Latour in principle can study endless number of chains, postphenomenologists seem to be restricted to two. [. . .] But the difference between the two approaches is more subtle than that, for in these short chains the postphenomenological perspective can bring to light things that remain invisible to actor-network theory. The postphenomenological perspective, for instance, offers a more nuanced look at the connections between the entities in its chains.

These two traditions and approaches can then be integrated into a three-dimensional view of social networks of human and non-human actants, such as that proposed by Contractor, Monge, and Leonardi (2011). According to these authors, although communication technologies and behaviors can be conceptually put on the same level, there are also differences in practice: "For example, although a person might decide to retrieve information from his or her friend on one day, and from a technology (like a database) on the next day, it is unlikely that he or she would ever consider the technology his or her friend" (685). They then distinguish between (1) unidimensional (unimodel and uniplex) networks, made up of a single type of nodes, usually human beings, and of a single type of edges; (2) unimodal and multiple networks, made up of a single type of nodes but by multiple types of edges; (3) multimodal uniplex networks, with more types of nodes and only one type of edges; (4) multidimensional, multimodal, and multiplex networks, in which there are different types of nodes and edges; (5) the properly multidimensional networks, in which the relations are identified not only within a set of nodes but also between different sets of nodes. Lastly, they add a level (6) applicable to all precedents, in which nodes and edges can be added or deleted over time. It is precisely this complexity, but also this passion for differences and limitations even in the similarities, which seems to be lacking in many considerations of Bruno Latour, even if it must be recognized that he has more recently taken a more cautious position on the subject. Speaking of actor-network theory, he says that "[t]his theory played a critical role in dissolving overly narrow notions of institution, in making it possible to follow the liaisons between humans and non-humans, and especially in transforming the notion of 'the social' and society into a general principle of free association" (Latour 2013, 64). And yet, he adds immediately afterward,

> We understand this now, this method has retained some of the limitations of critical thought: the vocabulary it offers is liberating, but too limited to distinguish the values to which the informants cling so doggedly. [. . .] A tool in the war against the distinction between force and reason, it risked succumbing in turn to the unification of all associations under the sole reign of the number of links established by those who have, as it were, "succeeded."
>
> (Ibid.)

Similarly, from an epistemological point of view, he proposes to "*insert a wedge* between two modes that have been amalgamated to each other," namely "reference" and "reproduction" (73–74). And interestingly enough, he refers to the same story of Borges's with which I started this part of the book: "Borges has clearly warned us against the dream of a map at full scale. The *gain* in knowledge [. . .] stems precisely from the fact that the map *in no way* resembles the territory." He speaks then of a "strange means of transportation whose continuous back-and-forth movement along a fragile cable [. . .] gradually charges the map with a minuscule portion of the territory and extracts from the territory a full charge of signs" (78–79). The question arises whether such strange means of transportation is not the hermeneutic circle (which looks more here like a hermeneutic pulley) that lacks in most of Latour's considerations on the digital.

Notes

1. https://en.wikipedia.org/wiki/LICRA_v._Yahoo! Accessed June 10, 2019.
2. This is a sentence by the Italian rapper Frankie Hi-NRG MC in the 1998 song "Quelli che benpensano." It paraphrases the Gospel of Matthew (19:30): "But many that are first, shall be last, and the last shall be first." Interestingly enough, the so-called Matthew Effect, which can be summarized by the adage "the richer get richer and the poor get poorer," is based on the parable of the talents or minas, another passage of the Gospel of Matthew (which appears also in the Gospel of Luke).
3. In this context, I am not considering other approaches, like Lev Manovich's cultural analytics or the digital social research developed at the Oxford Internet Institute.

4. D. Lazer and R. Kennedy, "What We Can Learn From the Epic Failure of Google Flu Trends." *Wired*, January 1, 2015. www.wired.com/2015/10/can-learn-epic-failure-google-flu-trends/. Accessed June 10, 2019.

5. K. Severson, "Butterballs or Cheese Balls: An Online Barometer." *New York Times*, November 25, 2009. www.nytimes.com/2009/11/26/dining/26search. html. Accessed June 10, 2019.

6. www.artefaktum.hu/en/Latour.htm. Accessed June 10, 2019.

7. In the French version of the text the sentence has been changed: "We do not have just the World Wide Web for materializing relations: we are in the middle of a material infrastructure that makes the work enormously easy for us, for us as sociologists of science, and that we could call the World Wide Lab."

8. https://en.wikipedia.org/wiki/Flatland. June 10, 2019.

9. In the already cited 1998 lecture, the perspective is rather different. He speaks of "traceability," but such traceability is still related to the idea of a separation between the real and the virtual, and to the idea that the virtual can empower our real existences. At the end of the talk, he says: "can it [the Virtual] produce new Virtualities into the Society? That is, can it end up producing a different Society? In the etymology of Virtuality there is the Word Virtue." Such a perspective is undoubtedly due to the time the lecture was delivered, long before the birth and the rising of the social Web.

10. It is Tarde himself (2012, 52) who says how the verb "to have" should be preferred to the verb "to be": "All philosophy hitherto has been based on the verb *Be*, the definition of which was the philosopher's stone, which all sought to discover. We may have said that, if it had been based on the verb *Have*, many sterile debates and fruitless intellectual exertions would have been avoided."

11. "Quali-quantitative" is, in fact, the term that has been used by those who have "put into practice" Latour's theories. See, for instance, the issue 6/188 (2014) of the French journal *Réseaux*, entitled *Méthodes digitales: Approches quali/quanti des données numériques* (Digital methods: quali/quantitative approaches of digital data).

12. C. Anderson, "The End of Theory: The Data Deluge Makes the Scientific Method Obsolete." *Wired*, June 23, 2008. www.wired.com/2008/06/pb-theory/. Accessed June 10, 2019.

13. Regarding their definition, I am content to give the one recently proposed by Holmes (2017, 7), according to which Big Data are "datasets that are large in both size and complexity, with which new techniques are required in order to extract useful information from them." According to Kitchin (2014), emerging literature describes Big Data as (1) exhaustive, (2) detailed in resolution, (3) relational, and (4) flexible.

14. S. Lohr, "The Origins of Big Data: An Etymological Detective Story." *New York Times*, February 1, 2013. http://bits.blogs.nytimes.com/2013/02/01/the-origins-of-big-dataan-etymological-detective-story/. Accessed June 10, 2019.

15. Rosenberg (2013) writes that the first occurrence in English of the term "data" reported by the Oxford English Dictionary is in 1646, in a treatise on theology that speaks of a "heap" of data. In the seventeenth century, the word is used in the technical sense that it had in Euclid, to indicate the amount of data in the mathematical problems, as opposed to the *quesita*, which were instead the quantities sought. In philosophy, data were the principles that could not be doubted, either by their self-evident nature or by convention. In theology, the term indicated the truths of the Scriptures given by God and therefore indisputable. The word became common in the eighteenth century but taking another, almost opposite sense: data are not the undisputed bases of an argument, but the results of an experiment. The element

of self-evidence here remains because data would do nothing other than to unveil (in the sense of the Heideggerian *aletheia*) or to present reality as such.

16. In Severo, Feredj, and Romele (2016), we have proposed to introduce the alternative term "soft data." The first feature of soft data is their availability on the Internet. Soft data mainly consists—but not only—of new types of web 2.0 data (Facebook, Twitter, RSS, et cetera). They can be big or small. Secondly, soft data are easily accessible and storable through application programming interfaces (APIs) or by scraping tailor-made scripts. Thirdly, they are bottom-up sources of information.

17. Latour (2010b, 15) writes that "Simondon has grasped that the ontological question could escape [. . .] the obsession with the distinction between subject and object and could be rather seen in terms of vectors." He thus overcomes that separation of principle between human beings and things of the world, including obviously the technologies. As Simondon himself says (2012, 125), "our research aims to show that human thought must establish a relationship of equality, without privilege, between technologies and human beings." On the following pages (126–127), he insists on the fact that the condition of "inclusion of technical objects into culture would be that human being is neither inferior nor superior to technical objects, that she can approach them and learn to know them by entertaining with them a relation of equality, of reciprocity of exchanges: a social relation in some way"— a true principle of Latourian symmetry, therefore. But Latour (2010b, 16) also recalls with reason that "Simondon remains a classic, obsessed with the original and future unity, deducing its modes from one another, in a way that might rather recall Hegel. [. . .] In the end, multirealism would be nothing else than a long detour to come back at the philosophy of being."

References

Beaude, B. 2014. *Les fins d'internet*. Limoges: FYP.

———. 2012. *Internet: changer l'espace, changer la société*. Limoges: FYP.

boyd, d. 2014. *It's Complicated: The Social Lives of Networked Teens*. New Haven: Yale University Press.

———. 2001. "Sexing the Internet: Reflections on the Role of Identification in Online Communities." Paper presented at *Sexualities, Media, Technologies*, University of Surrey, 21–22 June. www.danah.org/papers/SexingTheInternet. conference.pdf. Accesses June 10, 2019.

boyd, d., and Crawford, K. 2011. "Six Provocations for Big Data." *A Decade in Internet Time: Symposium on the Dynamics of the Internet and Society*. http:// dx.doi.org/10.2139/ssrn.1926431. Accessed June 10, 2019.

Bryant, R.E., Katz, R.H., and Lazowska, E.D. 2008. "Big-Data Computing: Creating Revolutionary Breakthroughs in Commerce, Science, and Society." https://cra.org/ccc/wp-content/uploads/sites/2/2015/05/Big_Data.pdf. Accesses June 10, 2019.

Cardon, D. 2008. "Le design de la visibilité." *Réseaux* 152(6): 93–137.

Contractor, N., Monge, P., and Leonardi, P.M. 2011. "Multidimensional Networks and the Dynamics of Sociomateriality: Bringing Technology Inside the Network." *International Journal of Communication* 5: 682–720.

Diebold, F.X. 2012. "A Personal Perspective on the Origin(s) and Development of 'Big Data': The Phenomenon, the Term, and the Discipline." PIER Working Paper. http://papers.ssrn.com/sol3/papers.cfm?abstract_id=2202843. Accessed June 10, 2019.

Donath, J., and boyd, d. 2004. "Public Display of Connection." *BT Technology Journal* 22(4): 71–82.

Foucault, M. 1995. *Discipline and Punish: The Birth of the Prison*. New York: Vintage Books.

Fuchs, C. 2014. *Digital Labour and Karl Marx*. London and New York: Routledge.

Ginsberg, J., Mohebbi, M.H., Patel, R.S., Brammer, L., Smolinski, M.S., and Brilliant, L. 2009. "Detecting Influenza Epidemics Using Search Engine Query Data." *Nature* 457(7232): 1012–1014.

Halpern, O. 2014. *Beautiful Data: A History of Vision and Reason Since 1945*. Durham: Duke University Press.

Harman, G. 2009. *Prince of Networks: Bruno Latour and Metaphysics*. Victoria: Re-press.

Holmes, D. 2017. *Big Data: A Very Short Introduction*. Oxford: Oxford University Press.

Ihde, D. 1990. *Technology and the Lifeworld: From Garden to Earth*. Bloomington: Indiana University Press.

Iliadis, A., and Russo, F. 2016. "Critical Data Studies: An Introduction." *Big Data & Society*, http://journals.sagepub.com/doi/abs/10.1177/2053951716674238. Accessed June 10, 2019.

Kitchin, R. 2014. *The Data Revolution: Big Data, Open Data, Data Infrastructures and Their Consequences*. London: Sage.

Latour, B. 2013. *An Inquiry into Modes of Existence*. Cambridge, MA: Harvard University Press.

———. 2010a. "Networks, Societies, Spheres: Reflection on an Actor-Network Theorist." Keynote Speech for the International Seminar on Network Theory: Network Multidimensionality in the Digital Age: Annenberg School for Communication and Journalism, Los Angeles, February 19, 2010. www.bruno-latour.fr/sites/default/files/121-CASTELLS-GB.pdf. Accessed June 10, 2019.

———. 2010b. "Prendre le pli des techniques." *Réseaux* 5(163): 11–31.

———. 2007. "Beware, Your Imagination Leaves Digital Traces." *Times Higher Literary Supplement*, April 6, 2007. www.bruno-latour.fr/sites/default/files/P-129-THES-GB.pdf. Accessed June 10, 2019.

———. 2005. *Reassembling the Social: An Introduction to Actor-Network Theory*. Oxford: Oxford University Press.

———. 1996. *Aramis, Or the Love of Technology*. Cambridge, MA: Harvard University Press.

———. 1993. *We Have Never Been Modern*. Cambridge, MA: Harvard University Press.

Latour, B., Jensen, P., Venturini, T., Grauwin, S., and Boullier, D. 2012. "The Whole Is Always Smaller Than Its Parts: A Digital Test of Gabriel Tarde's Monads." *The British Journal of Sociology* 63(4): 590–615.

Lazer, D., Kennedy, R., King, G., and Vespignani, A. 2014. "The Parable of Google Flu: Traps in Big Data Analysis." *Science* 343(6176): 1203–1205.

Lazer, D., Pentland, A., Adamic, L., Aral, S., Barabási, A.-L., Brewer, D., Christakis, N., Contractor, N., Fowler, J., Gutmann, M., Jebara, T., King, G., Macy, M., Roy, D., and Van Alstyne, M. 2009. "Computational Social Science." *Science* 323(5915): 721–723.

Marres, N. 2017. *Digital Sociology: The Reinvention of Sociological Research*. Cambridge: Polity Press.

———. 2015. "Why Map Issues? On Controversy Analysis as a Digital Method." *Science, Technology & Human Values* 40(5): 655–686.

Marres, G., and Gerlitz, C. 2015. "Interface Methods: Renegotiating Relations Between Digital Social Research, STS and Sociology." *The Sociological Review* 64(1): 21–46.

Marres, N., and Weltevrede, E. 2013. "Scraping the Social? Issues in Live Social Research." *Journal of Cultural Economy* 6(3): 313–335.

Mashey, J. 1998. "Big Data and the Next Wave Infrastress." In *Computer Science Division Seminar*. Berkeley: University of California.

Mayer-Schönberger, V., and Cukier, K. 2013. *Big Data: A Revolution That Will Transform How We Live, Work, and Think*. Boston and New York: Eamon Dolan/Houghton Mifflin Harcourt.

Pentland, A. 2014. *Social Physics: How Social Networks Can Make Us Smarter*. New York: Penguin Books.

Rogers, R. 2013. *Digital Methods*. Cambridge, MA: The MIT Press.

Romele, A., and Severo, M. 2016. "The Economy of the Digital Gift: From Socialism to Sociality Online." *Theory, Culture & Society* 33(5): 43–63.

Rosenberg, D. 2013. "Data Before Facts." In L. Gitelman (ed.). *Raw Data Is an Oxymoron*. Cambridge, MA: The MIT Press, 15–25.

Scholz, T. (ed.) 2012. *Digital Labor: The Internet as Playground and Factory*. London and New York: Routledge.

Severo, M., Feredj, A., and Romele, A. 2016. "Soft Data and Public Policy: Can Social Media Offer Alternatives to Official Statistics in Urban Policymaking?" *Policy & Internet* 8(3): 354–372.

Simondon, G. 2012. *Du mode d'existence des objets techniques*. Paris: Aubier.

———. 2005. *L'individuation à la lumière des notions de forme et d'information*. Grenoble: Millon.

Tarde, G. 2012. *Monadology and Sociology*. Victoria: Re-press.

Turkle, S. 1995. *Life on the Screen: Identity in the Age of the Internet*. New York: Touchstone.

Venturini, T., Baya Laffite, N., Cointet, J-P., Gray, I., Zabban, V., and De Pryck, K. 2014. "Three Maps and Three Misunderstandings: A Digital Mapping of Climate Diplomacy." *Big Data & Society*. http://journals.sagepub.com/doi/abs/10.1177/2053951714543804. Accessed June 10, 2019.

Venturini, T., Cardon, D., and Cointet, J-P. 2014. "Présentation." *Réseaux* 6(188): 9–21.

Venturini, T., and Latour, B. 2010. "The Social Fabric: Digital Traces and Quali-Quantitative Methods." https://medialab.sciencespo.fr/publications/Venturini_Latour-The_Social_Fabric.pdf. Accessed June 10, 2019.

Verbeek, P-P. 2005. *What Things Do: Philosophical Reflections on Technology, Agency, and Design*. University Park: Penn State University Press.

Weiss, S., and Indurkhya, N. 1997. *Predictive Data Mining: A Practical Guide*. Burlington: Morgan Kaufmann.

Winner, L. 1980. "Do Artifacts Have Politics?" *Daedalus* 109(1): 121–136.

Zhao, S., Grasmuck, S., and Martin, J. 2008. "Identity Construction on Facebook: Digital Empowerment in Anchored Relationships." *Computers in Human Behaviors* 24: 1816–1836.

Conclusion

Today, there is an emerging interest for the potential role that hermeneutics might have in reflecting technologies. For instance, Coeckelbergh and Reijers (2016) have investigated the narrative capacities of technologies, trying to overcome the gap between textual narratives and material technologies. They have proposed to see the "reading" or "configuration" process as reciprocal, in the sense that it goes from the humans to technology, but also the other way around. They have also explored two hermeneutic distinctions, one related to the activity or passivity of the technology in the configuration process, another which is concerned with the extent to which technological mediation abstracts from the realm of human action. More generally, I would say that we are facing the beginning of a sort of "hermeneutic turn" in philosophy of technology, which implies a renewed interest for language, imagination, symbolism, et cetera.

In the case of digital technologies, such an interest is even more evident, because as I have said, in the digital it is immediately and intrinsically a matter of language, signs, and symbols, although of a specific kind. In recent years, there has been an increasing use of the term "digital hermeneutics," but there has not been unanimity into the meaning attributed to it. Two main distinctions that characterize the approaches: (1) between "methodological" and "ontological" digital hermeneutics and (2) between data- and text-oriented perspectives. The shift from a methodological to an ontological attitude will be at the center of the next part of this book. Here, I am going to focus on the latter distinction. Methodological digital hermeneutics does not represent a unitary field. Some academics, usually with a background in linguistic or digital humanities, affirm or implicitly admit that digital hermeneutics is about computer-mediated interpretation and understanding of texts or texts corpora, or about a texts reading-inspired attitude toward elements of digitality such as the code.

Burnard (1998), for instance, brought the attention to the hermeneutic implications of text encoding. For him, in the end, all markup is interpretive. Moreover, "markup maps a (human) interpretation of the text into a set of codes on which computer processing can be performed. It thus enables us to record human interpretations in a mechanically shareable way." Using texts' interpretation as a paradigm rather than as a research goal, van Zundert (2016) has recently argued that digital hermeneutics must not be reduced to a post-processing of what remains after the automated process of curation, analysis, and visualization. He cites, among others, Ramsay (2010), who theorized the "hermeneutics of screwing around." The only solution to the information overload—"so many books, so little time" is Frank Zappa's phrasing which the author discusses at the beginning of the article—that we are facing because of digital technology, especially the Internet, is to find a purposively selective and subjective path through it. The hermeneutics of "screwing around" is, then, a "highly serendipitous journey replacing the ordered mannerism of conventional research" (7). For van Zunden, instead, digital hermeneutics should be thought of as an intimate engagement with digitality and the software itself.

If several researchers have theorized digital hermeneutics through the lenses of (digitalized and digital native) interpretation and understanding of texts, some others, usually with a background in digital sociology, argue that digital hermeneutics should be more widely concerned with all forms of digital data.[1] For me, this perspective is to be preferred because it reflects the same process of universalization/deregionalization that has been accomplished by classic hermeneutics at the beginning of the twentieth century, when Dilthey brought hermeneutics beyond its regional use. Hermeneutics strived with him for being the proper methodology for *all* human and social sciences. To use Dilthey terminology, understanding (*Verstehen*) in human and social sciences always depends on the interpretation of an expression (*Ausdruck*) of someone's lived experience (*Erlebnis*).

The explosion of digital traceability and the development of digital methods are in some way the (promise of) realization of this epistemological dream. At the time of Dilthey and throughout the twentieth century, we certainly observed an enormous growth of the "documentality" (Ferraris 2012). However, on a closer inspection, documents and traces still concerned few acts and (social) events such as births, marriages, deaths, fines, and criminal records. Today, mobile and wearable digital devices, along with the increasing presence of digital recording systems in the environment we live in, has made digital traceability to be a "total social fact." Despite all possible criticism, I believe that Latour had the merit of bringing forward the very issue of digital traceability. Of course, information and communication are still relevant aspects of the digital, yet I believe that today recording, registration, and keeping track represent the most appropriate paradigm for understanding the digital and its consequences. Actually, for Bachimont (2018, 14) recording is more than a metaphor, because it is at the core of the technical structure of the digital:

An inversion [between communication and recording] occurred when IP protocol-based technologies introduced packet switching: [. . .] whereas, until then, we used to communicate without recording, and the issue of recording was eventually raised after the communication, the IP protocol imposed to record in the form of packets first, in order to communicate these same packets in a second stage. Instead of a wave propagation routing a signal without having to record it, we now have a registration that we want to carry from one place to another.

Now, digital hermeneutics can aspire to universality as far as it aligns itself with digital traces and data, the presence of which is henceforth as great as the human and social reality, a "map" that finally *almost* corresponds to the "territory." I say "almost" because digital hermeneutics, as I have already described it in length in the course of this first part of this book, can be actually defined as a general problematization of the supposed homology between (social) reality and its digital representations. In the existing literature, such problematization is based on ontological, methodological, and technological considerations.

For instance, for Østerlund et al. (2016), the hermeneutical dimension of digital social sciences depends on the ontological structure of the "trace data" themselves. Classic hermeneutics was concerned with texts but not with their materiality—that is to say, the apparatus for gathering the texts. Digital hermeneutics must be rather focused on the sociomateriality, the details of the specific system technology that captured the data. Gerbaudo (2016) has proposed a "data hermeneutics" as a methodological reply to the anti-interpretative ideology of contemporary "dataism." For him, data hermeneutics should be based on two processes: (1) qualitative sampling procedures to reduce the size of (social media) datasets; (2) the development of a "close data reading" that may help interpretation in relation to individual narratives, dialogical motivations, and social worldviews. Van den Akker et al. (2011) defined digital hermeneutics as "the encounter of hermeneutics and web technology": "interpretation of information in a digital environment," the "main aim of which is to investigate the relation between the human interpretation process and web application supporting that interpretation process." In particular, the authors have proposed a technical solution called Agora to support users of the Rijksmuseum Amsterdam collection database in their contextual interpretation and understanding of single objects by allowing them to build around it "historically meaningful narratives."[2]

Despite the several differences in sensibilities and approaches, one might say that digital hermeneutics generally consists in unveiling the 'hot' side of the mostly 'cold,' that is, highly quantitative, patterns characterizing the contemporary "digital dataism." It is important to recall that Ricoeur tried to overcome the Gadamerian dichotomy between truth and method, so 'hot' and 'cold' in digital hermeneutics must not be considered as an alternative.

In his major work *Truth and Method* (2004), Gadamer introduced a distinction that played a fundamental role in the hermeneutic debate during the second half of the twentieth century. According to him, hermeneutics must be understood in contraposition to the scientific, methodological, and quantitative perspective that reduces truth to the analytic explanation of discrete facts. According to Ricoeur (1991, 70–74), the merit of Gadamer consists in having taken seriously the epistemological question of Dilthey while remaining within a Heideggerian framework. His concern, indeed, is to confront the Heideggerian concept of truth with the Diltheyan notion of method. And yet, "the question is to what extent the work deserves to be called *Truth AND Method*, and whether it ought not to be entitled instead *Truth OR Method*" (71). In other terms, Gadamer's distinction is in reality an alternative: "either we adopt the methodological attitude and lose the ontological density of the reality we study, or we adopt the attitude of truth and must then renounce the objectivity of the human sciences" (75).

For Ricoeur there are two ways of doing hermeneutics. There is the "short route," which is the one taken by an ontology of understanding à la Heidegger. It consists into breaking with any discussion of method and in carrying "itself directly to the level of an ontology of finite being there to recover understanding, no longer as a mode of knowledge, but rather as a mode of being. [. . .] One does not enter this ontology of understanding little by little [. . .]: one is transported there by a sudden reversal of the question" (Ricoeur 2004, 6). And there is the "long route," the one he has tried to travel, which rather consists in undertaking a detour through semiotics and semantics and aspiring to ontology as a very last degree of meditation. Digital hermeneutics can be understood as an actualization in the digital age of the Ricoeurian plea for the methods without losing touch with existence. Digital hermeneutics does not consist for me in opposing the 'hot' existence to the 'cold' digital methods and objects, but rather into articulating them, in making existence, preconceptions, and specific worldviews emerge from an internal analysis of the methods and the objects themselves.

In Romele, Severo, and Furia (2018, 9–13), we have explored digital hermeneutics 'in action,' through a work on the literature about of the use of Twitter data for analyzing political opinion. In particular, we have resorted to the Ricoeurian model of the triple *mimesis* to describe the process that goes from traces to data (prefiguration), from data to methods (configuration), and from methods to information (reconfiguration). We have demonstrated that political opinion is not just a neutral evidence emerging through empirical 'cold' analysis, because preconceptions about what political opinion is and previous choices/affordances (the researchers' competences, the funding/time at disposal, et cetera) have a relevant impact on the result.

In this context, I would like to conclude by developing some philosophical remarks on the notion of "digital trace," which represents, in my opinion, the hermeneutic alternative to the concept of semantic information as it has been developed by Floridi. Ontologically, the notion refers to an unbridgeable gap between reference and meaning, between reality and its digital representations. Technologically, digital traces precede both data and information. Finally, methodologically, the notion recalls the highly interpretative character of the digital traces' treatment and "emplotment" into databases, and eventually in visualizations and their multiple actualizations.

From an etymological point of view, the English term "trace" derives from the ancient French *tracier*, which in turn derives from the Latin *trahere*, "to pull." The English term "track" seems to derive instead from the Middle Low German *trecken* which, although the philologists still debate on the thing (some assert, for example, a common non-Indo European but Semitic root), has nothing in common with the Latin *trahere*. In German, the term used today is *Spur* which comes from the Old High German *spor*, which indicated the imprint of the feet. According to the Grimm brothers' dictionary, the term is closely related to *Spüren*, "to sense," that is, to identify a track, an imprint, or a clue and follow it. For this reason, when we talk about traces, we think of fingerprints, but also of clues and indexes. Trace also refers to the mark, the inscription and the recording—the term that derives from the Latin *recordari*, "to remember" and that still reverberates in the Italian *ricordare* or in the Spanish *recorder* and that literally means "restore" (*re-*) in the "heart" (*corde*).

Firstly, it is the notion of trace in general to tell us something about the way in which digital traces should also be understood. I am thinking in particular of the French phenomenology of authors such as Levinas, Derrida, and Ricoeur.

In "The Trace of the Other" (1986), Levinas states that the trace is not a sign like any other. For sure, every trace can be taken as a sign. Levinas proposes the examples of a detective examining the area where a crime took place, a hunter following the traces of game, and the historian discovering ancient civilizations. In these cases, he says that everything is arranged in an order, in a world, where each thing reveals another or is revealed in a function of another. But this is a derivative way of understanding the trace. In fact, when understood in the proper sense, a trace does not serve to give order to the world, but rather to disturb it: "[h] e who left traces in wiping out his traces did not mean to say or to do anything by the traces he left. He disturbed the order in an irreparable way" (357)—as if the trace of a crime should not be looked at from the detective's side, but from that of the criminal who left it despite herself.

But if this is the proper meaning of the trace, then the trace par excellence must refer to something (or someone) that is essentially inappropriable. For this reason, the philosopher writes, "[o]nly a being that transcends the world can leave a trace. A trace is a presence of that which properly speaking has never been there, of what is always past" (358). Levinas is thinking of the Other, of the radical alterity irreducible to us, of an absolute ontology irreducible to any epistemology, in short to what for him is ultimately God. And indeed, the text ends with the assertion that "[t]he revealed God [. . .] maintains all the infinity of his absence. [. . .] He shows himself only by his trace" (359). An interesting notion, that of the Levinassian trace, but also one that is paradoxical and essentially impossible. In other words, it is a concept that ends up condemning all kind of epistemology, to denounce epistemology as essentially wrong, not to say violent because, after all, every form of knowledge is also a reduction of otherness to the self.

Derrida understood this notion in a different manner. The article "The Trace of the Other" was originally published in 1963 in the journal *Tijdschrift voor Filosofie*, two years after the publication of *Totality and Infinity* and a year before the publication of the critical reading of this text by Derrida, "Violence and Metaphysics," appeared in the *Revue de métaphysique et de morale*. It seems to me that the criticism that Derrida addresses to Levinas in this text is analogous to the way in which Derrida himself thinks of the trace right from Levinas.

In "Violence and Metaphysics" (then published as part of *Writing and Difference*) Derrida wonders whether it is really possible to get rid of the Western metaphysical language and its violence, as Levinas would like. In other words, he is doubtful about the feasibility of disengaging from that Greek-Western conceptuality that reduces the otherness to its knowledge, and ethics to ontology. In fact, Levinas himself is obliged to use this conceptuality and this language to undermine them. Derrida writes about this (1978, 111): "We are not denouncing, here, an incoherence of language or a contradiction in the system [of Levinas]. We are wondering about the meaning of a necessity: the necessity of lodging oneself within traditional conceptuality in order to destroy it." For Derrida it is not so much a question of breaking with Western metaphysics; and, besides, breaking is a gesture that recalls the same violence with which metaphysics, from Aristotle to Heidegger, has separated the beings from the Being. Instead he suggests to follow that rupture, or rather that series of fractures that can be already found in Western metaphysics, in its language and its texts and to which Derrida refers to with the well-known term "différance."

It is the same kind of gesture that Derrida makes for the notion of trace. Where he speaks of traces, Levinas thinks of the definitive rupture with Western ontology and the ethical openness toward the absolutely Other. In *Of Grammatology* (2016), Derrida instead reports this rupture of the trace within the being itself—being as text. To put it bluntly, there is no need to go so far, to God or to the Other, to find an inappropriable otherness. Just look around us, in the world, in the daily language and in the multiplicity of its signs. Indeed, the structure of the sign itself is determined by the trace of an 'other' which will be forever absent. In short, his effort consists in 'secularizing' the Levinasian trace—and the Heideggerian ontological difference. For Derrida, the trace is a mark of the absence of presence, an always-existing absent present that characterizes the Being itself. The trace is a mark of that deferment of reference, of that differential relation between the signifier and the signified, and between the signs of a language, which the French philosopher borrows from Saussure and extremizes through Nietzsche. The problem with Derrida (or at least with his scholars) is that these insights are transformed into a general praise of uncertainty and of the principle of undecidability with respect to the question of meaning. Although 'secularized,' his attitude is not less paradoxical than that of Levinas, to the extent that its effects remain merely deconstructive and not very constructive. For this reason, I prefer the Ricoeurian perspective to that of Derrida. Ricoeur is interested in the notion of trace as well, but for him such a notion is the matrix of a difficult but still possible epistemology.

In the third volume of *Time and Narrative*, Ricoeur escapes to a ruinous alternative.[5] The first is that of simply understanding the trace as the mark and the effect of a cause. He (Ricoeur 1988, 118) criticizes for example "contemporary historiography, with its data banks, its use of computer and information theory." In particular, on the same pages, he says that "the data in a data bank is suddenly crowned with a halo of the same authority as the document cleansed by positivist criticism." It is an illusion that breaks the link with the past and with the lived experience in this past, which forgets "the debt to the dead." In other words, it is a relationship with the past (and, more generally, with everything that is absent) that cancels the distance between the representant and the represented person. On the other hand, however, Ricoeur also denies that the trace is pure significance, a sign of an all Other which would be by essence inaccessible. Discussing Levinas, he declares to share with him the idea that a trace is distinguished from all the signs that get organized into systems, because it disarranges such order. And yet, he prefers the idea of a "relative" and "historical" Other (125). For Ricoeur, the trace is therefore the matrix of a difficult but possible epistemology. An encounter between Heidegger's existential and ontological time—"a fundamental time of Care, the temporality directed toward the future and toward death," writes Ricoeur (120)—and the worldly and ontic one. Ricoeur develops his notion of trace within the context of a specific epistemological problem, which is that of historiographic knowledge. Think about how much difference there is with Heidegger, who reduces every question of this kind to the radical questioning about the historicity of

Dasein. The trace remains for Ricoeur an epistemological issue, which shows a certain way of making history and, more generally, human and social sciences. Ontological aspects are not absent, but they are reached just at the very end of a long epistemological detour.

It seems to me that Paul Ricoeur's hermeneutics of the trace encounters the evidential paradigm elaborated by the historian Carlo Ginzburg in the text "Clues: An Evidential Paradigm" (1989). Incidentally, the term used in the Italian version of the article is not "clues," as in the English one, neither *traces*, as in the French version, but *spie*, which means both "spies" and "indicators" or "warning lights." In the same way, the word used is not "evidential," which recalls the proof and something "obvious to the eye or mind," but *indiziario*, *indiciaire* in French indicates a hint or a clue—nothing evident or obvious then.

Ginzburg speaks, for example, of the "Morellian method," the methodology introduced by the Italian art critic Giovanni Morelli to distinguish original paintings from their copies based on minor details such as earlobes, fingernails, shapes of fingers and of toes. Imitating the great traits of a painter, like Leonardo's smiles, is easy, but it is more difficult (also because the imitator does not pay much attention to them) to imitate minor and secondary details. Ginzburg then speaks of Sherlock Holmes—the comparison between Morelli and the Conan Doyle character was first proposed by Enrico Castelnuovo—and Freud, who in his essay on the *Moses* of Michelangelo, recognizes his debt to Morelli. But more interesting is that Ginzburg traces the origin of the paradigm further back, long before the birth of all sorts of scientific epistemology:

> Man has been a hunter for thousands of years. In the course of countless chases he learns to reconstruct the shapes and movements of his invisible prey from tracks on the ground, broken branches, excrement, tufts of hair, entangled feathers, stagnating odors. He learned to sniff out, record, interpret, and classify such infinitesimal traces as trails of spittle. He learned how to execute complex mental operations with lightning speed, in the depth of a forest or in a prairie with its hidden dangers.
>
> (102)

A hermeneutic of the trace would therefore be much wider (both in depth and width) than the classic hermeneutics of texts, documents, or monuments. And if it goes back such a long way, why could it not move forward, toward a digital hermeneutics? Ginzburg refers, for example, to the Hippocratic medicine, whose method is based on the practice of detecting, interpreting, and understanding symptoms—he speaks of developing histories of individual diseases. For the Italian historian, the evidential paradigm is perhaps what all the human and social sciences have in common, and which, as just mentioned, has very deep roots in human history.

However, Ginzburg's limit consists in having proposed—in this sense, not differently from Ricoeur—a clear-cut distinction between the Galilean universal and abstract paradigm and evidential paradigm. This would be, for him, at the roots of the distinction between 'hard' and 'soft' sciences. Louis Liebenberg (1990) went in a certain sense further because he considered tracking as the paradigm of *all* sciences. The hunter thinks just like a physicist (who, among other things, studies the traces of the particles she does not see), by observation and induction. In both cases, there is indeed a structuring and knowledge of some patterns (always adaptable and guided by the initial conditions and preconceptions) starting from sense data. The trackers develop, of course, a savoir faire, but also a real theoretical knowledge that allows them to distinguish the species between them and the individuals within the same species. The author speaks of a "systematic tracking," of general models that are enriched or are refuted through experience and thanks to collective dynamics reminiscent of those of scientific communities.

Regarding digital traces more specifically, I believe that the notion helps in recalling that there is no science, at least among data sciences, which is not interpretative to some extent. It also helps in bringing the "ascetism of meaning" back to a "semantic of things." Jean-Marc Ferry (2007, 83) resorted to the Peircian distinction between index, icon, and symbol to say that Ginzburg's aspiration to found human and social sciences on a grammar of natural signs denies the specificity of the symbolic grammar that characterizes most of them. In fact, according to Peirce, unlike the symbol, which maintains a merely conventional link with the thing signified, and unlike the icon, in which there is a bond of similarity, the index is characterized by physical contiguity with the reference. Early examples proposed by Peirce, such as those of the relation between the wind and the weathercock or between the murderer and his or her victim, suggest the idea of a true material causality.

However, I would say that it is precisely the function of the notion of trace to question the supposedly symbolic ascetic grammar of data sciences. The notion of trace, in general, is a theoretical tool supporting the critique of the idealism of matter that I have been discussing in the overture. The concept of digital traces in particular brings us back to the material aspects of the digital and might be used for contrasting the illusion of transparency in the field—see Krämer (2012, 2–3) about traces in general and Serres (2002, 1) about digital traces on this point.

The concept of digital traces has a sort of paradoxical nature. Indeed, as it has been said, if the traces are absence of presence, and the digital's pretention is of being a "map" as extended as the "territory," then, properly speaking, there can be no digital traces at all. As the philosopher François-David Sebbah writes in this regard (2015, 123.), "in a sense, it [the digital trace] 'presentifies' more than any other type of traces; one might believe that it saves the ghosts of the past better than any other;

but, by saving them too much, it consumes them." In other words, being the digital trace too perfect, it also ends up losing its primary function. Similarly, Collomb (2016) has argued that "trace" and "digital" are two contradictory words: if in fact the trace à la Derrida refers to a temporality that always exceeds the pure presence, digital traces-based predicting profiling practices aim instead at compressing temporality, since the past and the present are used to anticipate the future.[4]

In their introduction to issue 2 of the *International Journal of Communication* "Digital Traces in Context," Hepp, Breiter, and Friemel (2018, 443) have rightly used the notion of digital trace to criticize the illusion of homology which can be found, for instance, in Latour's texts on the digital that I have been discussing: "[S]uch approach misunderstands digital traces as something 'neutral' offering us 'direct access' to the social world. However, digital traces [. . .] rely on the technical procedures of governing institutions that actively produce this kind of information." One might say that the aim of a digital hermeneutics consists precisely in returning digital traces to their dimension of traces. It means, in short, to show the existence and resistance of a gap between the reality and its digital representations, to recall, as I have already said, that "the virtual never ended." Ginzburg's evidential paradigm cannot be the point of departure for a critical epistemology of the digital because it does not account for the digital's pretentions of homology. Rather, it must be considered as the point of arrival, the result of a hermeneutic deconstruction of the digital's dominant worldview itself. The notion of digital trace thus helps us in undertaking such "backward questioning" of the digital, that goes back from information to data, and from data to digital traces. If one applies a diaphorical definition of data to the digital, as Floridi does, he or she will have already forgotten the material/historical origins and the symbolic context in which the digital epistemology itself is always-already entangled.

Notes

1. It must be said that even in this case texts interpretation often remains a paradigm or a metaphor through which data and data analytics are understood.
2. Armaselu and Jones (2016) analyzed the users' responses to the different visualization forms offered by Transviewer, an XML-TEI-based platform allowing the exploration of historical documents. Armaselu and van den Heuvel (2017) have recently considered how interpretation is supported and shaped by metaphors embedded in an interface. Their article is based on the analysis of three uses of the z-text model and Z-editor interface that allows the user to create and explore zoomable texts.
3. On this point see O. Abel, "The Trace as Answer and Question." http://olivi erabel.fr/ricoeur/the-trace-as-answer-and-as-question.php. Accessed June 10, 2019.
4. Against the idea of such a perfection of the digital traces, and the digital as the perfect archive and recording system, see for instance Nielsen (2018), who

resorts to web archiving practices for arguing that the Web is more "an ever-changing scene characterized by mutability and unpredictability as the technical infrastructure, the applications running on the Web, and the content change and evolve rapidly" (51).

References

Armaselu, F., and Jones, C. 2016. "Towards a Digital Hermeneutics? Interpreting the User's Response to a Visualisation Platform for Historical Documents." www.dhbenelux.org/wp-content/uploads/2016/05/106_Armaselu-Jones_FinalAbstract_DHBenelux_long.pdf. Accessed June 10, 2019.

Armaselu, F., and van den Heuvel, C. (2017). "Metaphors in Digital Hermeneutics: Zooming Through Literary, Didactic and Historical Representations of Imaginary and Existing Cities." *Digital Humanities Quarterly* 11(3). www.digitalhumanities.org/dhq/vol/11/3/000337/000337.html. Accessed June 10, 2019.

Bachimont, B. 2018. "Between Formats and Data: When Communication Becomes Recording." In A. Romele, and E. Terrone (eds.). *Towards a Philosophy of Digital Media*. London: Palgrave Macmillan, 13–30.

Burnard, L. 1998. "On the Hermeneutic Implications of Text Coding." http://users.ox.ac.uk/~lou/wip/herman.htm. Accessed June 10, 2019.

Coeckelbergh, M., and Reijers, W. 2016. "Narrative Technologies: A Philosophical Investigation of the Narrative Capacities of Technologies by Using Ricoeur's Narrative Theory." *Human Studies* 39(3): 325–346.

Collomb, C. 2016. "Et s'il n'y avait pas de traces numériques?" www.academia.edu/3510979/Et_sil_ny_avait_pas_de_traces_num%C3%A9riques_. Accessed June 10, 2019.

Derrida, J. 2016. *Of Grammatology*. Baltimore: John Hopkins University Press.

———. 1978. *Writing and Difference*. Chicago: The University of Chicago Press.

Ferraris, M. 2012. *Documentality: Why It Is Necessary to Leave Traces*. New York: Fordham University Press.

Ferry, J-M. 2007. "Le paradigm indiciaire." In D. Thouard (ed.). *L'interprétation des indices. Enquête sur le paradigme indiciaire avec Carlo Ginzburg*. Lille: Septentrion, 91–101.

Gadamer, H.G. 2004. *Truth and Method*. London and New York: Continuum.

Gerbaudo, P. 2016. "From Data Analytics to Data Hermeneutics: Online Political Discussions, Digital Methods and the Continuing Relevance of Interpretative Approaches." *Digital Culture & Society* 2(2): 95–112.

Ginzburg, C. 1989. "Clues: Roots of an Evidential Paradigm." In *Clues, Myths, and the Historical Method*. Baltimore: John Hopkins University Press, 87–113.

Hepp, A., Breiter, A., and Friemel, T.N. 2018. "Digital Traces in Context. An Introduction." *International Journal of Communication* 12: 439–449.

Krämer, S. 2012. "Qu'est-ce donc qu'une trace, et quelle est sa fonction épistémologique? État des lieux." *Trivium: Revue franco-allemand de sciences humaines et sociales* 10. https://journals.openedition.org/trivium/4171. Accessed June 10, 2019.

Levinas, E. 1986. "The Trace of the Other." In M.C. Taylor (ed.). *Deconstruction in Context: Literature and Philosophy*. Chicago: The University of Chicago Press, 345–359.

Liebenberg, L. 1990. *The Art of Tracking: The Origin of Science*. Claremont: David Philip.

Østerlund, C.S., Crowston, K., and Jackson, C.B. 2016. "The Hermeneutics of Trace Data: Building an Apparatus." IFIP Working Group 8.2 Working Conference, Dublin. https://citsci.syr.edu/sites/crowston.syr.edu/files/Crowston_Osterlund_Jackson_The_Hermeneutics_of_Trace_Data-Full_Paper.pdf. Accessed June 10, 2019.

Nielsen, J. 2018. "Recording the Web." In A. Romele, and E. Terrone (eds.). *Towards a Philosophy of Digital Media*. London: Palgrave Macmillan, 51–76.

Ramsay, S. 2010. "The Hermeneutics of Screwing Around: Or What You Do with a Million of Books." www.leeannhunter.com/digital/wp-content/uploads/2014/08/RamsayBooks.pdf. Accessed June 10, 2019.

Ricoeur, P. 2004. *The Conflict of Interpretations: Essays in Hermeneutics*. London and New York: Continuum.

———. 1991. *From Text to Action: Essays in Hermeneutics, II*. Evanston: Northwestern University Press.

———. 1988. *Time and Narrative, Volume 3*. Chicago: The University of Chicago Press.

Romele, A., Severo, M., and Furia, P. 2018. "Digital Hermeneutics: From Interpreting with Machines to Interpretational Machines." *AI & Society*, online first, 1–16.

Sebbah, F.D. 2015. "Traces numériques: plus ou moins de fantômes?" In C. Larsonneur, A. Regnauld, and P. Cassou-Nougès (eds.). *Le sujet digital*. Dijon: Les presses du réel, 114–127.

Serres, A. 2002. "Quelle(s) problématique(s) de la trace?" https://archivesic.ccsd.cnrs.fr/sic_00001397/document. Accessed June 10, 2019.

Van den Akker, C., van Erp, M., Aroyo, L. Segers, R., van der Meij, L., van Ossenbruggen, J. et al. 2011. "Digital Hermeneutics: Agora and the Online Understanding of Cultural Heritage." ACM Web Science Conference, Koblenz. www.cs.vu.nl/~guus/papers/Akker11a.pdf. Accessed June 10, 2019.

van Zundert, J.J. 2016. "Screwmeneutics and Hermenumericals. The Computationality of Hermeneutics." In S. Schreibman, R. Siemens, and J. Unsworth (eds.). *A Companion to Digital Humanities*. Oxford: Blackwell Publishing, 331–347.

Part 2

*E*magination

Introduction

In the previous part of this book, digital hermeneutics was treated from an epistemological and methodological perspective. In this second part, I propose to take a step forward. My intention is to perform, within the context of digital hermeneutics, the same ontological turn that classic hermeneutics accomplished in the twentieth century. As already mentioned, Dilthey universalized hermeneutics by electing it to the role of methodology for *all* human and social sciences. But Heidegger brought hermeneutics to another level. For him, hermeneutics has very little to do with texts' or cultural objects' interpretation; rather, it concerns the ontological-existential conditions that make all interpretations and understandings possible. From this perspective, hermeneutics deals with interpretation and understanding as the proper way human beings act and interact in the world. In other words, hermeneutics is a general meditation on human beings as 'interpreting animals.' The main question of this part of the book is if, and eventually to which extent, it is possible to attribute to some digital machines, or at least to an emerging part of them, interpretational capacities. To put it differently, can we say that in the digital we are increasingly dealing with 'interpretational machines'?

One of the greatest risks in using hermeneutics in such a context is that classic hermeneutics, both in its epistemological and ontological version, has been radically anthropocentric. Consider the case of Dilthey: in attempting to expand hermeneutics, and to give human sciences an autonomous methodology with respect to that of the natural sciences, he actually regionalized and limited it in its possible developments. It was, in short, dignification by means of excessive specification. It is also important to underline how this enclosure not only affects hermeneutics or human sciences, but also, and perhaps above all, the field that is supposed to be foreign to them: the 'nature' of the natural sciences. If interpretation and understanding concern exclusively human things, then research on nature is suddenly deprived of all quests about sense and meaning.

In his personal diary, on February 15, 1861, Dilthey writes: "only what is human is understandable to man; we understand all the rest only by analogy with the human" (GS 7, 225).[1] In 1875, he states: "Our proper world is society, not nature. Nature is mute for us, and it is only from time to time that a glimpse of life and interiority passes through it thanks to the power of our imagination" (G 5, 61). Again, in 1911, following an observation by Schleiermacher, he affirmed: "Interpretation would be impossible if the vital externalizations were entirely foreign. It would be useless if there were nothing foreign to them. It is therefore located between these two opposite extremes. It is necessary wherever there is something extraneous of which the art of understanding must [and can] appropriate" (GS 7, 278).

In the Diltheyan perspective, we understand the behavior or action of another person only if we are able to attribute an intentionality to her, and therefore a psychic and inner life. Nature is not understandable for him precisely because it has no interiority and has no real intentions. For this reason, according to Dilthey, "understanding of nature—*interpretatio naturae*" is just a figurative expression.

The Heidegger before the *Kehre* transformed hermeneutics into a philosophy, and more precisely into an ontology of existence. The 'local,' methodological, and epistemological hermeneutics are only derivatives of this most fundamental hermeneutics, which designates nothing less than the being of *Dasein*: "To the extent that this hermeneutic elaborates the historicity of Da-sein ontologically as the ontic condition of the possibility of the discipline of history, it contains the roots of what can be called 'hermeneutics' only in a derivative sense: the methodology of the historical humanistic disciplines" (Heidegger 1996, 33). Now, it is precisely by radicalizing the ontological dimension—that is, the privileged relationship of *Dasein* with the question of Being—of hermeneutics that Heidegger cannot but exclude all beings, except that being, the *Dasein*, which poses the question about Being from hermeneutics itself. Heidegger famously stated that the animal is "poor in world" (*weltarm*). In fact, hermeneutics is not for him just a matter of more or less conscious preconceptions and interests that guide all actions and perceptions in the world. In this sense, it is clear that animals have their "environment" (*Umwelt*) as well. Heidegger's hermeneutics is also, and above all, a matter of a privileged relationship of the *Dasein* with the question of Being.

Let us consider Jackob von Uexküll's reflections about the *Umwelt* of the adult female tick. After mating, the female climbs to the tip of a twig on some bush. There she can stay for a long time, at a slow pace of life and without the need for any sustenance, at such a height that she can drop upon small mammals that may run under her or be brushed off by larger animals. The approaching prey is not revealed by sight (the tick is blind and deaf) but by her sense of smell. As explained by von Uexküll (1992, 321), "[t]he odor of butyric acid, that emanates from the skin glands of all mammals, acts on the tick as a signal to leave her watch-tower and hurl herself downwards." If she falls on something warm—the tick has "a fine sense of temperature"—then she will have to find a hair-less spot to feed herself; otherwise, she will have to climb on the bush again. The tick blood repast is also her last meal: she will drop to earth, lay her eggs and die. The tick's *Umwelt* is very simple then, but also particularly effective: "The whole rich world around the tick shrinks and changes into a scanty framework consisting, in essence, of three receptor cues and three effector cues—her *Umwelt*. But the very poverty of this world guarantees the unfailing certainty of her action, and security is more important that wealth" (325).

From these considerations, von Uexküll draws the idea that interpreta-tion is the way in which every living organism interacts with its environ-ment. The functional circle of the tick is based on a selection of signs (or rather, signals, *Merkmale* or *Merkzeichen*, "bearers of meaning"): there is, therefore, no perception of the world that can be said to be neutral. It is through this principle, for example, that biosemiotics extends its analysis to inter-species interpretative relations and to interpretative or proto-interpretative functions at the biochemical level.

But Heidegger interprets these same analyses in a completely different way. It is in the 1929/1930 seminar *The Fundamental Concepts of Meta-physics* that Heidegger speaks, for the animal, of a "poverty in world" (*Weltarmut*). On the one hand, he emphasizes a great closeness between human being and animal. He praises the work of von Uexküll ("the most fruitful thing that philosophy can adopt from the biology domi-nant today"), in particular for having discovered "the relational structure between the animal and its environment" (Heidegger 1995, 263). But the philosophical limit of this work is for Heidegger not to account for the difference between the ways in which humans and animals inhabit their respective worlds: a difference that "is not simply a question of qualita-tive otherness" and "especially not a question of quantitative distinction in range, depth, and breadth." It is rather about "whether the animal can apprehend something as (*als*) something, something as a being, at all. If it can not, then the animal is separated from man by an abyss" (264). It is the *Als-Struktur*, "as if structure," which is explained in § 32 of *Being and Time*, where Heidegger speaks of "understanding and interpreta-tion." Understanding something *as* something means not only seeing a thing in its context of use and significance. It also, and above all, means understanding this very structure and, without leaving the hermeneutic circle, opening up to possibilities for renegotiation and reconfiguration of one's relationship with the world.

In the 1929/1930 seminar, Heidegger then proposed a distinction between the behavior (*Benhemen*) of animals and human comportment (*Verhalten*). The difference between humans and animals consists in the fact that the former are able to decide their own conduct, and for this reason, enjoy a certain degree of freedom. "The behavior of the animal," writes Heidegger (1995, 237), "is not a doing and acting, as in human comportment, but a driven performing (*Treiben*)." Agamben (2003, 57) recalls that in his *Parmenides*, a course delivered in 1942–1943, Heidegger criticizes Rilke, who in the eighth Duino Elegy talks about the animal (*die Kreatur*) as what sees the open "with all its eyes," in contrast to human being, whose eyes have been "turned backward" and placed "like traps" around him. While human beings always have the world before them, the animal instead moves in the open, in a "nowhere without the no." Obviously Heidegger does not agree with Rilke, because true openness, for him, is the openness of Being that questions us, and not the openness of the 'wordly world.' Although Heidegger does not express himself in these terms, even here it is possible to say that the idea of an *interpretatio naturae* is a figurative way of understanding things. In this case, however, the genitive is no longer to be understood in its objective sense but in its subjective sense.

Heidegger, however much he says that human beings have no essence and that this coincides at the bottom with existence, however much he tries to determine and describe categories of life (the existentials) that do not determine a priori life, establishes a difference between human and animal world that is a difference in principle, and that by principle excludes all that is non-human from the realm of authenticity.

The history of hermeneutic anthropocentrism can be brought up to the present day. One could speak, for example, of Gadamer who in "Citizens of Two Worlds" (1992) uses the Kantian distinction between theoretical and practical reason in order to differentiate between "concepts of nature" and "concepts of freedom." And one could refer to even more recent publications, such as the book by Johann Michel, *Homo Interpretans: Towards a Transformation of Hermeneutics* (2019), in which interpretation is reduced to a very specific series of human activities and competences. But during the second half of the twentieth century, as I will show in the conclusion of this second part, some researchers made the effort to go beyond this perspective, while remaining within a hermeneutic framework. In particular, some authors have been interested in the question of nature's hermeneutics. I will propose to 'translate' the idea developed in this context, especially in the emerging fields of environmental hermeneutics and biohermeneutics, into the debate about digital technologies. The aim of the next two chapters is to lay the foundation for such an 'import.' In order to do this, I am going to focus on the issue of productive imagination, which recalls the processes of schematization and interpretation of the world in which human and non-human actions and interactions with the world are always engaged.

In the *Critique of Pure Reason*, Kant (1998, 272) presents the transcendental schematism as the bridge between understanding (categories) and sensibility (intuition, appearance): "there must be a third thing, which must stand in homogeneity with the category on the one hand and the appearance on the other, and makes possible the application of the former to the latter. This mediating representation [. . .] is the *transcendental schema*." Such a schema must arise from a faculty that is prior to both sensibility and understanding, which Kant calls the productive or transcendental imagination. The term used by Kant is *Einbildungskraft*, which refers to the image (*Bild*), but above all to the fact of constructing (*bilden*) or synthesizing within the process of objectifying.

However, the main references in the next chapters will be not Kant, but thinkers such as Ricoeur and Simondon, who proposed an externalized (and, in the case of Ricoeur, also semioticized and historicized) version of productive imagination. For Ricoeur, the schematism is not a "hidden art in the depths of the human soul," as argued by Kant (273). The synthesis between receptivity and spontaneity happens outside, in linguistic expressions such as symbols, signs, metaphors, and narrations. Moreover, the French philosopher does not exclude, at least in principle, that schematism can take place in and through other materialities other than linguistic concretizations. Simondon has described imagination as a form of practical and technical schematism. The scheme is not a mental entity for him, but rather an operation that is made through or even by the things themselves. On the one hand, it is my intention to 'exalt' the interpretational emerging capacities of digital machines, which in some cases support human schematizations yet in others replace them. On the other hand, I also want to 'frustrate' some human pretentions in terms of imagination, creativity, and ultimately freedom. It is precisely the passage through exteriority (technology, but also history, culture, society, and the body) that should bring us to temper some of our expectations in this respect.

Note

1. The work of Dilthey is cited according to the *Gesammelte Schriften* (GS; Dilthey 1957–2006).

References

Agamben, G. 2003. *The Open: Man and Animal*. Stanford: Stanford University Press.

Dilthey, W. 1957–2006. *Gesammelte Schriften*. Stuttgart and Göttingen: Teubner Verlag/Vandenhoeck & Ruprecht.

Gadamer, H.-G. 1992. "Citizens of Two Worlds." In D. Misgeld, and G. Nicholson (eds.). *Hans-Georg Gadamer on Education, Poetry, and History. Applied Hermeneutics*. Albany: SUNY Press, 209–20.

Heidegger, M. 1996. *Being and Time*. Albany: SUNY Press.

———. 1995. *The Fundamental Concepts of Metaphysics*. Bloomington: Indiana University Press.

Kant, I. 1998. *Critique of Pure Reason*. Cambridge: Cambridge University Press.
Michel, J. 2019. *Homo Interpretans. Towards a Transformation of Hermeneutics*. Lanham: Rowman & Littlefield.
Von Uexküll, J. 1992. "A Stroll Through the Worlds of Animals and Men: A Picture Book of Invisible Worlds." *Semiotica* 89(4): 319–391.

3   Imaginative Machines

In the previous part of the book, I have insisted on the fact that the 'new' of new technologies does not have much to do with information and communication, but rather with the exponential growth of the capacities of recording and keeping track. Here I will rather argue that the digital as it is today does not concern just recording and hence memory, but also imagination, and more specifically what philosophers, from Kant onward, have called productive imagination. For me, emagination (electronic or digital imagination) as I call it, does not contradict "e-memory" (Clowes 2013), but is an emerging property of it.

The hypothesis of this section is threefold. First, it will be argued that human productive imagination is always-already externalized. Digital technologies are one of the privileged places where the schematism takes place nowadays. Second, it will be said that some digital technologies imitate with increasing fidelity the way human productive imagination works. Third, it will be affirmed that in the age of Big Data and new algorithmic machines, digital imagination is taking the upper hand on human imagination, or at least a different path. Callon and Latour's principle of symmetry, according to which humans and non-humans (artifacts, nature, institutions, et cetera) should be integrated in the same conceptual framework and assigned an equal amount of agency, was based on a sort of *epoché*, a suspension of any kind of judgment and consideration about the intrinsic properties of the social actants. In this section, an integration of such a principle is somehow proposed: humans and certain machines should also be placed within the same conceptual and agency framework because they share the transcendental capacity of productive imagination.

When does interpretation begin? For several authors, interpretation is a sort of exception, which supervenes in the event of non-understanding or misunderstanding—and whenever there is the genuine will, of course, to understand something which is for some reason alien to us (a text, a cultural object, but also another living being, et cetera). But in most cases, our relationship with the world is oriented by the immediateness of habits. For me, as I have already argued in section 1.1, one should rather distinguish between 'white' or 'dead' interpretations and 'living' interpretations. Habits are precisely this: forms of 'dead' interpretations which sometimes are very convenient to keep this way. Imagine, for instance, having to question every morning the nature of the relationship with your partner, children, and job. Yet there are cases in which such immediate relationships with the world are and need to be questioned, and their interpretational nature is suddenly revitalized. I would even argue that there is no relation with the world which is not interpretative to some extent. Hans Lenk (1995) has distinguished for instance among six different degrees of interpretation: (1) proto-interpretation (*Urinterpretation*), which consists of primary schematizations, biologic, genetic, and invariable, like the fact that one can distinguish between bright and dark; (2) pattern-interpretation (*Musterinterpretation*), that allows recognition of the similitudes and the equivalencies among forms, colors, and so on; (3) conventional conceptualization (*konventionelle Begriffsbildung*), which is socially established and culturally transmitted. It can be pre-linguistic or linguistic; (4) classification-interpretation (*Einordnungsinterpretation*), that is, the conscious process of classifying elements in structures and schemes that can be linguistically described; (5) justification-interpretation (*Rechtfertigungsinterpretation*), which implies justificatory and theoretical reflections on the classifications proposed; (6) meta-interpretation (*Metainterpretation*), that is, the interpretation of the interpretations. Thus, one can affirm that 'interpretation is said in many ways,' but that there is no *essential* difference between the different meanings of the term.

This chapter will be developed in three steps. Firstly (3.1), I will deal with Ricoeur's theory of narrative, which is also a theory of productive imagination. According to the French philosopher, narrative is made of two elements: (1) the threefold *mimesis* (prefiguration, configuration, and refiguration), which is a movement of appropriation and distantiation on the basis of a process of representation of human action, and (2) the *mythos* (emplotment, *mise en intrigue*), that is, the capacity of giving coherence to the heterogeneous elements of a story. *Mimesis* and *mythos* are present in both fictional and non-fictional, historic and (auto) biographical stories. Once understood in the light of narrative, productive imagination has to do with two things: first of all, with the capacity, which is also a possibility, an individual and collective right, and an ethical task, of accounting for oneself before the others by means of the stories someone tells about himself or herself. In addition, one might say that narrative has to do with accounting for the others by means of the stories that someone tells about them. Second, it regards the fact that all stories, written or read, heard or told, by me or the others, about me or the others, fictional or not, have a heuristic force on human existences.

Secondly (3.2), it will be argued that human imagination is always-already extended. It will be also shown how *mimesis* and *mythos* are precisely the way the digital works. The *mimesis* is there insofar as the digital always produces dynamic representations of the world, which interpret the world, and can in their turn be interpreted. The *mythos* is present if we accept Lev Manovich's (2013a) thesis according to which "there is only software," and that "software = database + algorithms." For Kant (1998, 193–194), "thoughts without content are empty, intuitions without concepts are blind." Likewise, it will be said that algorithms without database are empty, database without algorithms are blind.

Third (3.3), the specificity of Big Data and new algorithmic machines, compared with older digital machines, will be introduced and discussed. Today, Big Data and new algorithmic machines represent the promise of giving our actions and existences a meaning that we are incapable of perceiving, for lack of sensibility (data) and understanding (algorithms). In the Web 1.0, digital productive imagination was, so to say, still below human imagination; in the social Web, there is rather a correspondence, or at least a distance which is close to zero. In the Web of today, the relation between human and digital imagination is going to be inversed, since the latter is overpassing the possibilities of the former. Or at least, even without wanting to confront them, it seems fair to say that digital imagination is taking an autonomous path which has concrete consequences on our decisions.

3.1 Productive Imagination

Ricoeur has never explicitly devoted an entire work to the issue of imagination. However, many scholars have noticed the relevance that this notion had in his thought (Kearney 1998; Taylor 2006; Foessel 2007; Amalric 2014). In particular, this section is going to focus on the role that productive imagination has played, often in the background, during the 1970s and the 1980s, when the French philosopher quit the domain of the symbols in favor of metaphors and narrations.

In order to enter the discussion, a peripheral point is taken: a debate Ricoeur had with Cornelius Castoriadis during a radio broadcast on the French radio channel France Culture in 1985. This debate is illustrative of the way Ricoeur understands productive imagination as the "golden mean" between the reproductive and the creative imagination. According to Ricoeur, "the idea of an absolute novelty is inconceivable. There is new only as a rupture with the old: there is a pre-settled (*pre-réglé*) before us that we unsettle in order to settle it differently" (Castoriadis and Ricoeur 2016, 44); "we are never in a sort of shift from nothing to something, but from something to something, from other to other—that goes from the configured to the configured, never from the formless to the form" (46).

In these passages, the French philosopher is taking position against his interlocutor. Indeed, Castoriadis advocates for a radical imaginary constitution, that is, for the possibility of absolute novelty in history and society by means of imagination: "The self-institution of society implies that we always work in an already settled context, manipulating and modifying the rules; *but also laying down new rules, creating them. This is our autonomy*" (44. Emphasis mine). Castoriadis opposes himself with respect to the tenet *ex nihilo nihil fit*, "out of nothing comes nothing." For him, there is a radical imagination, "a social instituting imaginary." He uses the term "radical" to oppose this imagination "to the 'secondary' imagination which is either reproductive or simply combinatory (and usually both). [. . .] It is radical because it creates *ex nihilo* (not *in nihilo* or *cum nihilo*)" (Castoriadis 1994, 138).

Ricoeur is rather equidistant from a creative and a reproductive concept of imagination, represented respectively for him by Jean-Paul Sartre and David Hume. In the most illuminating reflections he has devoted to this topic, he disposes the main philosophical theories of imagination on two axes: on the side of the object, the axis of presence (that is, imagination as the trace of presence, a weakened form of it) and absence (that is, imagination as the other-than-present, dream, hallucination, et cetera); on the side of the subject, which is not going to be considered here, the axis of fascinated consciousness (that is, zero critical consciousness, the image is confused with real, as it is the case for the power of lies and errors denounced by Blaise Pascal), and critical consciousness (that is, critical distance is fully conscious of itself, and imagination is an instrument for the radical critique of the real) (Ricoeur 1991, 170–171).

On the side of the object, Ricoeur argues that to imagine, individually or collectively, does not have to do with creating *ex nihilo*, but rather with reorganizing the elements of what already exists. Imagination has a double valence with respect to reference: on the one hand, it is directed elsewhere, that is, it is a form of suspension or distanciation from reality. On the other hand, reality remains its origin and scope. Sooner or later, imagination sets its sights on reality again, following what Ricoeur calls a "new reference-effect," which corresponds to the power of imagination of "reconfiguring" reality. For him, all imaginative act, that is, all inventions, creations, and interpretations, are done by "standing on the shoulders of giants." Incidentally, the metaphor of dwarfs standing on the shoulders of giants (*nanos gigantum humeris insidentes*), attributed to Bernard of Chartres and today made famous by Google Scholar, is for me particularly representative of the current digital structure and culture.

Let us consider the case of fictional novels. In the second volume of *Time and Narrative* (1986), the French philosopher insists on the fact that experimental novels such as those of Robbe-Grillet's *nouveau roman* do not completely break from Aristotelian and biblical tradition, based mainly on emplotment and the fact that a story must have a beginning and an end. For him, these are just specific kinds of narration in which the emplotment and the sense of an ending are close to zero, but not completely abolished. The fact is that all possible transgressions of the classic narrative model rest on a knowledge, and hence still a form of belonging to the model and the tradition itself. For instance, in the case of the *nouveau roman* the emplotment is not erased, but the task of it is simply displaced from the author to the reader, who must schematize by herself. Consider another example, those David Lynch's movies in which the coherence of the story offered to the spectator is always partial, precisely because is the spectator herself that must 'schematize.' As a consequence, all schematizations or synthesis of the events and the scenes in the movies will be partial and never conclusive. Many of you have certainly spent hours discussing with friends about the 'true' and 'right' interpretation to give to *Mulholland Drive* without ever reaching a final agreement. Ricoeur would say that in order to transgress the classic cinematic model of Hollywood cultural industry in which the spectator passively receives an already-schematized product, Lynch must have a certain knowledge of it, its history, theory, and techniques. Hence, Lynch's way of producing cinema does not abolish the model, but rather transforms it.

This approach suggests that freedom has less to do with liberation than with the renegotiation of frontiers with the social and technical milieus which surround us. I will develop this point below. Notice also how this perspective fits with our age, in which cultural productions, along with our own identities, are often the result of the recombination of elements borrowed from other cultures and epochs. This is all the more true considering the way text, sound, and video contents circulate and are reconfigured on the Web 2.0.[1]

To come back to the purposes of this section, it is important to stress that this possibility of recombining the already-existing elements of reality depends, in turn, on a movement of distanciation from reality and reappropriation of it. Such a movement, however, does not take place in interiority (*in interiore homine*), as many philosophers, from Augustine to Descartes and Kant, have said. Rather, distanciation and appropriation are made possible by forms of externalization and concretization of the human language, such as symbols, metaphors, and narratives. In the course of his work, Ricoeur has extended his analysis of productive imagination from the unity of the word (symbol) and the sentence (metaphor) to that of the text as a whole (narrative). It is precisely in his analysis of the configurative function of narrative that he most explicitly identified the role of productive imagination (Kearney 1998, 162). The model he refers to is the way Aristotle links, in the *Poetics*, the mimetic function of the tragedy to the mythic structure that the tragic poet has constructed. The impact of this model goes beyond the limits of fictional stories, because for Ricoeur the Aristotelian theory of tragedy is the paradigm of all forms of narrative. Moreover, a relevant part of our identities is constituted narratively (Ricoeur 1992, 113–168).

The French philosopher has developed his theory of the threefold *mimesis* in the last volume of *Time and Narrative*. The first *mimesis* (prefiguration) has to do with the predisposition of human action to be told and textualized. In fact, prefiguration deals with the constitution of a plot as imitation of human action. In order to do this, it is necessary to recognize the semantic of the action (that is, the structure distinguishing it from the physical movements), the symbolic of the action (that is, the signs, norms, and rules that mediate it), and the narrative of the action (that is, the temporal elements that make it capable of being narrated) (Ricoeur 1988, 54). The second *mimesis* (configuration) corresponds to the emplotment: it is the moment in which the different textualized elements of the human action are combined and recombined according to a specific temporal coherence. Ricoeur defines the *mythos* as "the temporal synthesis of the heterogeneous" or as a "discordant concordance" (66). As argued by Richard Kearney (162), "by narrative configuration he [Ricoeur] means [. . .] the ability to create a plot which transforms a sequence of events into a story. This consists of 'grasping together' the individual incidents, characters and actions so as to compose a unified temporal whole." Such a narrative synthesis enacts what for Kant is the productive power of the transcendental imagination.

Yet, narrative is never *for its own sake*, for its own glory. Language and all its forms always have a reference that is external to the language itself. The possibility of emplotment, indeed, is rooted in human concrete actions. The same emplotment is always-already on the way back to the world. The third *mimesis* (refiguration) has precisely to do with the application of the text to the world of the hearer or the reader (71). To put it slightly differently, refiguration represents the heuristic value of narrative, its "performativity." This term is actually misleading, since it refers to the codified effects of speech and language. Narrative, instead, opens up to a non-codified, and hence, potentially risky, novelty. Narratives, at least the good ones, always have unintended consequences on their readers and hearers, but also on their own authors. When we write, read, and hear a story, we are confronted with the actions and choices of its characters, we judge them, and in this way, we end up judging our own life. For this reason, Ricoeur (1992, 115) has affirmed that literature is "a vast laboratory in which we experiment with estimations, evaluations, and judgements of approval and condemnation through which narrativity serves as a propaedeutic to ethics."[2]

To sum up what has been said so far, I would argue that Ricoeur's contributions to the theory of imagination are three. Firstly, there is a moderation of some enthusiasms about human creativity and genius. In the transcription of a series of lectures given at the Centre de Recherches Phénoménologiques in Paris between 1973 and 1974, he affirms that it is precisely in the figure of the genius that one can find at the same time the realization and failure of the Kantian project: the realization, because the genius is one who invents without imitating; the failure, because without imitation, that is without outer traces, externalized both in time and space, imagination is reduced to the sphere of subjectivity, to "a pleasure experienced in the internal play of the faculties, as in a dance that goes nowhere" (Ricoeur 2002, 49). No specific cognitive function, but also no 'worldly' function is ascribable to the imagination of the genius. "It follows," writes Ricoeur, "that universality does not result from the participation of everyone in a beautiful thing, but from the essential communicability of a fundamentally subjective judgment" (Ibid.). The reasons for this failure lie, for the French philosopher, in Kant's misunderstand of the original meaning of the Aristotelian *mimesis*, and of what for Aristotle is the pleasure of "learning the genre." Kant distinguished between genius and imitation because he did not possess a good concept of imitation, such as the Aristotelian one. In fact, the Aristotelian *mimesis* already implies "a moment of creation, since the tragedy imitates the actions of real human beings recreating them in a *mythos*, which is at once a fiction, a fable and a coherent composition, an intrigue endowed with its own logic" (Ibid.).

Secondly, Ricoeur makes of schematism something that happens in the exteriority of linguistic forms such as symbols, metaphors, and narratives. As for the metaphor, for instance, he considers it as having a semantic rather than cognitive nature. The innovative power of metaphorical imagination lies in the linguistic possibility of establishing similarities in dissimilarities. Such possibility does not concern the single word, but rather the entire sentence and, in some cases, as in poetic language, it can involve an entire written discourse. Metaphors establish an acceptable deviation from ordinary predication; hence, they are particular (insofar as they are sort of voluntary "category mistakes") forms of schematization, articulation of heterogeneous elements according to an innovative, and yet still understandable and acceptable, coherence compared to the reality of the standard predication.[3]

Thirdly, productive imagination and schematism no longer concern only the knowledge of nature, but apply to the arts as to the techniques/technologies, to both scientific and cultural interpretation and understanding. In the Newtonian perspective of Kant, we are within a system that is always the same despite the differences of context—culture, tradition, society, et cetera. Instead, for Ricoeur, the schematism is situated. If it were not for the excessive naturalism implicit in this analogy, I would borrow from Lenk (1995) the idea according to which schematism resembles the formation of neuronal ensembles that evolve and adapt over time to our practices and needs. Another inspiration, always in the context of the several hermeneutic reappropriations of the Kantian schematism, might be Rudolph Makkreel, who has integrated the fixity of the first *Critique* with the dynamism of the third one: "the *Critique of Judgement* does not simply rely on the fixed rules and archetypes of the first *Critique*. The principle of reflective judgement is adaptive to the particular contents of experience and articulates order through the mutual adjustment of parts and wholes" (Makkreel 1990, 154). In short, he is talking of a schematism and therefore a productive imagination that not only applies to actions and interactions with the world in general, but also in which the categories tend to change as the conditions of sensibility change.

3.2 From Imagination to Emagination

The goal of this section is to apply the notion of productive imagination to the context of digital technologies. My hypothesis is that imagining is not exclusive to humans or animals. Rather, one can think of productive imagination as being also at work in some digital technologies. It is precisely this productive imagination that I call emagination. To be more precise, emagination means two things: first, that these machines play a fundamental role in human schematization. Productive imagination never takes place in the hidden depth of human soul, but always depends—sometimes partially, sometimes entirely—on the technologies we have at our disposal. Such a statement is part of a wider perspective according to which "to understand consciousness in humans and animals, we must look not inward, into the recesses of our inside; rather, we need to look to the ways in which each of us, as a whole animal, carries on the process of living in and with and in response to the world around us" (Noë 2009, 7). In this sense, using tools is one of the simplest ways in which we extend beyond the limits of our brains and bodies.

Second, emagination refers to the fact that the digital can partially account for a human faculty that is notorious for its obscurity. In fact, Kant's entire sentence is: "this schematism of our understanding [. . .] is an art concealed in the depth of the human soul, *whose true operations we can divine from nature and lay unveiled before our eyes only with difficulty*" (1998, 273. Emphasis mine).

Certainly a risky territory is entered: using digital computers for understanding human mind could bring to identify human mind and digital computers, to represent the mind as an information-processing system. Yet, it is sure that "computers can't think on their own any more than hammers can pound in nails on their own. [. . .] For this reason, we make no progress in trying to understand how brains think by supposing that they are computers" (Noë 2009, 169). And yet it is possible to use digital technologies as material metaphors for understanding some aspects of the human mind. The approximation of productive imagination and the digital is in my opinion legitimate insofar as it remains within the limits of an analogical thinking—remember the Simondonian notion of "transduction" I have introduced in the conclusion of the first part of the book.

It is noteworthy that Ricoeur has used his threefold *mimesis* in the field of architecture, which is a technique and an art of material building. According to him (2016, 32–33), architecture is analog to narrative, because in architecture we pass from a stage of "prefiguration," which is related to the act of *inhabiting*, to a second stage, the act of *building* (corresponding to the narrative emplotment), to reach a final stage of "refiguration," which is "the rereading of our towns and all our dwelling places." Concerning the architectural prefiguration, the French philosopher affirms that even before any architectural project, human beings have built somehow because they have inhabited. Architectural configuration is "a spatial synthesis of the heterogeneous," because architectural projects consist of articulating independent variables (units of space, massive forms, boundary surfaces) in an adequate unity. In architectural reconfiguration, Ricoeur sees an inhabiting "as a response, even as an *answer* to building, on the model of the agonistic act of reading, because it will not suffice for an architectural project to be well thought-out, or even for it to be held to be rational for it to be understood and accepted" (Ricoeur 2016, 39).

It is also interesting that, in *The Rule of Metaphor* (2004, 283), Ricoeur compares his theory of metaphor to Max Black's theory of models: "metaphor is to poetic language what the model is to scientific language. In scientific language, the model is essentially a heuristic instrument that seeks, by means of fiction [. . .] to lay the way for a new, more adequate, interpretation." This is noteworthy insofar as the first two levels of Black's hierarchy of models, two out of three, are occupied by technoscientific models: scale models (for instance, the model of a ship, the enlargement of something very small like an oxygen atom or the DNA helix, et cetera) and analog models (such as hydraulic models of economic systems or the use of electrical circuits in computer). This means that, at least in principle, Ricoeur does not exclude that schematism can take place elsewhere than in language.

Two caveats, however, are necessary here. First, when it comes to Black, his scale of models is a scale of abstraction. The last level is occupied by theoretical models, and the example he gives is Maxwell's representation of an electrical field in terms of the properties of an imaginary incompressible fluid. Second, with regard to Ricoeur, he continues to refer to language as the paradigm for all kind of synthetization. According to the intentions of this book, instead, language is just one of the possible external places in which schematism can happen.

Several approaches have proliferated in the last three decades that conceptualized the human mind and its activity not only as an extended, but also as embodied, enactive, embedded, distributed, and situated (Thompson 2007). Despite differences and nuances, they all share a critique of the traditional cognitive sciences that considered cognition in abstraction from both the body and its natural, social, and technological environment. This is the case, for instance, of the well-known extended mind hypothesis. "Where does the mind stop and the rest of the world begin?" is the opening question of Andy Clark and David Chalmers's famous article (1998). They both take position against the "realists," according to whom the demarcation of skin and skull are also the limits of the mind, and against the "idealists" who believe that the exteriority of meaning carries over into an externalism of mind. They advocate what they call an "active externalism," based on the active role of the environment in driving cognitive processes.

The example they begin with to show "the tendency of human reasoners to lean heavily on environmental supports" (8) is that of a person sitting in front of a computer screen. In all other cases they account for, first and foremost that of Alzheimer sufferer Otto's notebook, the doubt remains as to whether the coupled system (Otto's mind + his notebook) is not too skewed in favor of Otto's mind.[4] Indeed, the existence of the notebook's contents entirely depends on Otto's moments of lucidity. However, the coupled system starts to find its equilibrium only once the issue is no longer a mere recording but the synthesis is *actively* and *autonomously* operated by the technological artifact. This is precisely what happens with computers. Furthermore, once that quantity and capacity of synthesis start to become important, the equilibrium begins to skew in favor of the digital artifact. This is not a critique of the extended mind hypothesis per se. In fact, the intention of Clark and Chalmers is not so much to study the interaction between two cognitive or proto-cognitive systems, as indeed the term "coupling," especially bearing in mind the Husserlian *Paarung*, could suggest. Rather, it is for them a matter of seeing how some objects, which by themselves are not thought of as either cognitive or non-cognitive, become part of some cognitive routine (Clark 2011, 87). For me, it is rather a matter of showing how a new form of complex mediation, no longer reducible to a single cognitive system, arises in some of these cases.

In what more specifically concerns imagination, the idea according to which human productive imagination is always-already extended into technologies has been developed by Bernard Stiegler in the third volume of *Technics and Time* (2001), through the notion of tertiary retention. Taking the melody as an example, Edmund Husserl affirmed that the present of perception is in fact a matter of primary retention. In order to perceive sounds as a unique melody, the previous note must be retained when hearing the present note and must anticipate the note to come. Husserl distinguished this retention from the secondary retention, which is memory properly speaking, such as when we remember a melody we have heard in the past. According to Stiegler, however, Husserl's limitation consists in not having attributed any kind of relevance to the memory made of artifacts and mnemotechnical technologies. Through a fascinating meta-reading of the *Critique of Pure Reason*, he argues that Kant has not only confused primary and secondary retention, but also wrongly attributed tertiary retention to the unity of apperception:

> Yet, it would be impossible to avoid noticing that Kant's own flux of consciousness [. . .] manufactures and constitutes itself in its unity in the course of his writing of the various works constituting his *oeuvre*. [. . .] A situation such as this [. . .] is possible only because the imagination's primary and secondary syntheses are essentially synthesizable through this synthetic flux (of consciousness) constituting an "objective memory" analogous to a book or film.
>
> (Stiegler 2001, 46)

More recently, Yuk Hui has extensively applied the theories of Stiegler to the domain of digital objects. He fully developed, for instance, the concept of "tertiary protention" for what concerns Big Data analysis: "Through the analysis of data [. . .] the machines are able to produce surprises (not just crises) by identifying a possible (and probable) 'future,' a specific conception of time and space that is always already ahead but that we have not yet projected" (Hui 2016, 241).

Stiegler's theses are interesting not just because he specifically refers to imagination, but also because he has a critical attitude toward the issue of human technological extensions and hybridizations. Consider, for instance, what Clark (2011, 103–104) says concerning Kim Sterelny's objection about the extended mind hypothesis. For Sterelny, epistemic artefacts such as Otto's notebook operate in a common and often contested space, that is, a shared space apt for sabotage and deception by other agents. But according to Clark, in the end such critique reinforces the extended mind hypothesis, insofar as it shows how "in some contexts signals routed via perceptual systems are treated *in the way more typical of internal channels*," that is, as if they were transparent and reliable. I would say that Clark is right and wrong at the same time. He is right, because it is actually true that human beings tend by nature to resort to external cognitive artifacts as if they were part of their cognitive system. But he is also wrong, because this does not change the fact that such extensions can be manipulated, and that this phenomenon cannot be neglected by a (critical) philosophy of the extended mind nor by a (critical) philosophy of technology. It is as if Clark was so focused on accounting for the process through which artifacts become part of the cognitive process and hence " 'transparent' in use" (37), that he does not really care about some important social, ethical, and political consequences of our technological mediations and hybridizations.[5]

As I have said, Stiegler's perspective in this respect is more interesting. For him, it is not just a matter of preferring the mnemotechniques to the mnemotechnologies, that is, the good old intentional acts of externalizing memory and imagination against the new, often digital artifacts that take control of and automate people's memory and imaginative practices (Smolicki 2018). On this point, Clark (2011, 40) is right when he criticizes an anonymous referee according to which "cognitive enhancement requires that the cognitive operations of the resource be intelligible to the agent." If this was the case, then all good cognitive enhancement would be also irrelevant. For Stiegler, it is rather a matter of developing a pharmacology of the mnemotechnologies, and of the digital in particular, that is to say an analysis of the digital both as a poison and a possible remedy. The problem, for him, is not that psychic individuation is always-already externalized and collectivized. This is a fact that has to do with our human nature. Rather, the problem is that through digital technologies, such externalization and collectivization can also produce forms of "de-individuation"—one might even say "alienation." However, another digital is possible that encourages what he calls a "deliberative transindividuation": "the academic community should be at the origin of these annotation processes, which should produce the spaces for the conflicts of interpretation and scientific controversies essential for a rational implementation of what is now called the open source, open science, social innovation, etc." (Stiegler 2014, 26).

In this chapter, I propose to go further than Stiegler: first, by trying to describe, through the notions of *mimesis* and *mythos*, how productive imagination works; second, by saying that in the case of the digital it is not just a matter of externalization of the human productive imagination. Digital technologies, I would say, imitates with increasing fidelity the way human productive imagination actually works. Digital technologies, or at least an emerging part of them, are then "imaginative machines." This means that digital technologies work by *mimesis* and *mythos*. *Mimesis*, as it has been shown, is a process of distanciation and appropriation. And *mimesis* is present in digital technologies insofar as they depend on a process of dynamic representation of reality.

In an illuminating article, Yoni Van Den Eede has summarized and systematized what he called the theories of technology transparency: Martin Heidegger's, Marshall McLuhan's, Latour's, Don Ihde and Peter-Paul Verbeek's postphenomenology, and Andrew Feenberg's critical theory of technology. According to him, despite the different perspectives of these approaches, they all refer more or less explicitly to transparency as the fact that "something is not perceived, that is 'invisible,' or escapes conscious attention—it still *is* there in some capacity, but one sees 'through' it" (Van Den Eede 2010, 154). In continuity with these theories, here it is affirmed that, despite their apparent immediacy, digital technologies are hermeneutic in nature.

Ihde (1990, 72–112) has famously distinguished four kinds of technologically mediated I-world relations: (1) embodied relations, whose specificity lies in the high transparency of the technological artifact after a period of adaptation (for instance, glasses); (2) hermeneutic relations, which give a representation of the world that interprets the world, and that must in its turn be interpreted (for example, thermometers and maps); (3) alterity relations, in which the relation with the world is temporarily suspended, and the technology itself assumes the role of interlocutor/competitor (for instance, computer games); and (4) background relations, when a technology creates the conditions of our own relation with the world (for example, heating and lighting systems).[6]

It is evident that digital technologies have something to do with all these relations. The mobility and portability of objects such as smartphones, smartwatches, and smartbands are the preamble to a radical embodiment of digital technologies. The fact that we trust online services like Google Maps for getting information demonstrates that we consider more and more digital technologies as valuable interlocutors. And of course, the increasing presence of digital sensors in the environment we live in (room, house, city, et cetera), along with the fact that such a "sensibility" augments the environment's capacity to adapt to our needs/intentions, shows how digital technologies are capable of establishing background relations between the world and us. And yet all the digitally mediated I-world relations are based on the hermeneutic possibility of translating—"transcoding" may be a better term—the world in a digital representation.

To take just one example, the Internet of Things (IoT) is commonly considered the paradigm of the integration between online and offline, the being-online of a naturally offline object such as an electrical appliance, a farm animal, a human heart, et cetera. But in order to be integrated into the Internet, all objects need an IP address as a unique identifier. This is the reason why the IoT will have to use the internet protocol version 6 instead of the version 4, which routes most Internet traffic today, but allows 'just' 4.3 billion unique addresses. Hence, each object must have its own symbolic representation—in this case, a numerical label—in order to be recognized as being part of the online network.

Digital technologies are always based on a mostly invisible process of symbolic distanciation from the world, and it is only on this basis that they become effective in the world. Marianna van den Boomen has opportunely defined digital objects like the desktop icons as "material metaphors." With this term, she meant double entry signs, since they involve two kinds of code: machine-readable digital code, to which the icon refers indexically (as already said, in Peirce terminology, an index is a sign entertaining a relation of contiguity with its object), and human-readable code, to which the icon refers symbolically (for Peirce, a symbol is a sign denoting an object by mere convention and abstraction), for example, as mail, file, or program (van den Boomen 2014, 40). In other words, digital technologies always pass through a representation that, thanks to its twofold nature, symbolical and indexical, is also performative. What is interesting in this perspective on digital technologies is that it goes beyond the alternative between representativity and performativity. Indeed, it is the case only insofar as digital technologies represent the world that they can also recombine (plus give people a certain room for maneuvering) its elements. Furthermore, this approach goes beyond the presumed end of the virtual we have discussed in the first part of the book, since the possibility of seizing the real for digital technologies always depends on a certain distance they take from it.

The *mythos*, the emplotment of digital technologies, is founded on such a hermeneutic movement of distanciation and appropriation. Their logic is analog to the emplotment it has been introduced above. According to Lev Manovich (2013a, 9), the software is behind all discussions concerning digital technologies:

> If we limit critical discussions of digital culture to the notions of "open access," 'peer production,' 'cyber,' 'digital,' 'Internet,' 'networks,' 'new media,' or 'social media,' we will never be able to get to what is behind new representational and communication media and to understand what it really is and what it does. If we don't address software itself, we are in danger of always dealing with its effects rather than the causes.

The software is for Manovich the basis of all digital expressions, from the creation and sharing of cultural and social artifacts (for instance, a video on YouTube, a page of Wikipedia, a comment on Facebook, et cetera) to the way private enterprises and public institutions (GAFA, NSA, Microsoft Research, MIT media lab, and so on) use the digital traces left by the prosumers (producers and consumers) for purposes of marketing, surveillance, research, et cetera. The analogy with narrative emplotment consists in the fact that the logic of software depends on the articulation of two elements: data structures and algorithms. As Manovich says, "*a medium as simulated in software is a combination of a data structure and set of algorithms.* [. . .] We have arrived at a definition of a software 'medium,' which can be written in this way: Medium = algorithms + a data structure" (207). For him, "to make an analogy with language, we can compare data structures to nouns and algorithms to verbs. To make an analogy with logic, we can compare them to subjects and predicates" (211).

As far as Manovich privileges software, and the articulation within the software between data structures and algorithms, he remains a sort of material idealist—more specifically, an idealist material dualist. Indeed, the specificity of data structures is that they do not include the data itself. For this reason, I propose to 'translate' Manovich's data structures into databases.[7] To make an analogy with the productive imagination, one could compare then databases to sensibility and algorithms to the forms of understanding. Indeed, the function of an algorithm consists of reorganizing data according to a certain coherence. For instance, this is the case with the main data processing task performed by computers, which is sorting, that is, putting a collection of values into order (Hill 2016). And this is also the case with all other performances operated by digital tools: an image on Instagram is a collection of pixels on which different filters have been applied; an Excel document of scientometric data downloaded from Scopus is a database on which we can apply different algorithms for data treatment and visualization such as Gephi (Romele and Severo 2016); from a user perspective, the Web itself is a database on which algorithms like Google PageRank operate.

According to Kant (1998, 193–194), "thoughts without content are empty, intuitions without concepts are blind." Likewise, one might say that algorithms without database are empty, database without algorithms is blind. It is maybe this "imitation game" that provokes the fascination that we have for digital technologies and makes them the symbolic form of our age. "Software is the interface to our imagination and the world," said Manovich (2013b), because it "has replaced a diverse array of physical, mechanical, and electronic technologies used before the twenty-first century to create, store, distribute and access cultural artifacts, and communicate with other people." But the truth is that digital technologies are not only interfaces to our imagination and the world but are one of the ways (probably the main one today) in which productive imagination externalizes and realizes itself in the world.

3.3 Emagination, Big Data, and New Algorithmic Machines

The argument that has been developed until now is twofold. On the one hand, it has been argued that digital technologies act 'imitating' the human productive imagination. On the other hand, it has been said that the human productive imagination takes place in digital technologies. These two perspectives are not contradictory, but they can integrate each other: human productive imagination operates outside, in digital technologies (but also, of course, in other media technologies), and digital technologies work through a schematism that is analogous to that of human beings. We have a circle in front of us (hermeneutical, of course), on top of which another circle is inserted.

Nevertheless, this image is still simplistic: first, because a digital technology is never an isolated system. In its creation and use, it rather depends upon a complex interaction between technological affordances, social conditions, and cultural assumptions. Digital technologies are sociotechnical systems; that is to say, they are entangled with a material infrastructure and a social/cultural superstructure. The extended productive imagination is never 'mine,' and it is for this reason always-already exposed to the risks of control, domination, et cetera. Theodor Adorno and Max Horkheimer have highlighted the relation between transcendental schematism and cultural industry. In fact, what they said about cinema is all the more true of digital technologies: "Kant said there was a secret mechanism in the soul which prepared direct intuitions in such a way that they could be fitted into the system of pure reason. But today this secret has been deciphered" (Adorno and Horkheimer 2003, 149). However, their perspective cannot be shared here, because they used to consider technology as a source of alienation per se. Second, it is simplistic because even in the case of self-tracking technologies (Van Den Eede 2015), the data collected and emplotted do not affect only the person these data are about.

In Rodighiero and Romele (forthcoming), we have argued that, however powerful, the model of the threefold *mimesis* risks also to be reductive. Indeed, the hermeneutic circle of data visualization is, so to speak, a *double* hermeneutic circle. The hermeneutic circle of the 'reader' must always be paired with the circle of the designer. Digital traces are collected and, through an opportune treatment, transformed into data and emplotted into databases. The designer will then resort to programming or digital tools to emplot data again in a visual form using a scope that renders them more understandable, and eventually aesthetically appreciable. The design process can lead to different forms of actualizations, of which computer screens are just the most common option. The production and collection of new digital traces will bring into question these actualizations, and the hermeneutic circle of the designer will start again. The hermeneutic circle of data visualization has, thus, the form of the lemniscate, or the infinity symbol (see Figure 3.1):

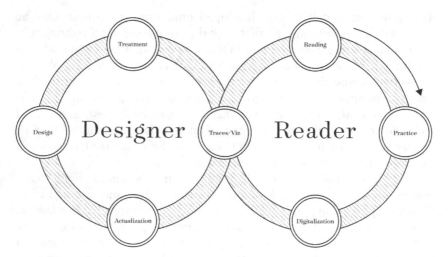

Figure 3.1 The Double Hermeneutic Circle of Data Visualization

Source: Rodighiero and Romele, forthcoming

Moreover, the Ricoeurian model of triple *mimesis* suggests that there is *one* author, *one* book, and *one* reader at a time. It is worth noting that even when Ricoeur writes of more than one author (e.g. the Bible), more than one book (e.g. the New Testament), and more than one reader (e.g. the public reading and listening practices in the Christian churches), he implicitly applies the classic model of solitary, silent writing and reading.

The Affinity Map, our case study in that article, clearly shows a collective engagement and collectivization in both sides of the hermeneutic circle.[8] Digital traces do not concern one unique, isolated individual. Even when digital traces are collected for quantifying individuals, they are continuously used as touchstones for comparison with someone else as a condition of visual proximity. The digital emplotment implies a series of stratified processes and technology uses that are the result of a mediation between human and non-human double intentionality—not to mention the fact that the designer does not work in solitude; she is usually part of a team, in continuous contact with clients and the audience in a process of dialogic design.

In this context, I would like to focus on a second element, which concerns specifically an emerging part of the digital machines. The frontier between imaginative and non-imaginative machines has not been considered yet. Luciano Floridi and James Sanders (2004) argued that agenthood relies on (1) interactivity, that is, the response to stimulus by change of state; (2) autonomy, which consists in the ability to change state without stimulus; (3) adaptability, which is the capacity to change the 'transition rules' by which state is changed. These are not universal properties but depend on a Level of Abstraction (LoA), the point of view from which a certain thing/situation is observed. In the case of artificial agents such as machine learning algorithms (the author proposes the example of MENACE, an "analogic" engine that learns how to play at Noughts and Crosses), we must consider them as moral agents (although aresponsible) at a certain LoA: not that of the single game, nor that of the code, but the LoA of the tournament. At this specific LoA, machine learning algorithms are not passive recording systems, nor do they seem to have a preprogrammed capacity of combination and recombination. Rather, they show an emerging adaptability, an emerging capacity of creative synthetization of the heterogeneous or what I have called *emagination*.

My hypothesis is that most of the algorithmic machines we deal with as users today can be said to be imaginative at a certain LoA. For example, the first time someone has to do with Facebook's Year in Review feature, she or he will probably have the impression of an authentic emplotment. The reiterated observation or use of the tool, however, will make its deterministic character appear. But things are in fact a little more complicated—first, because users usually do not have access to the code; second, because in complex machine learning and pattern recognition algorithms, evidence for a conclusion/decision is most of the time inscrutable, even for those who have implemented the algorithm (Mittel-stadt et al. 2016, 4). As Norbert Wiener (1964, 21–22), speaking about learning machines, pointed out, "[i]t may be said that all this unexpected intelligence of the machine has been built into it by its designer and pro-grammer. This is true in one sense, but it need not be true that all of the new habits of the machine have been explicitly foreseen by him." One can certainly imagine that there is always a higher LoA that makes these machines' choices appear as deterministic. Yet such an LoA might not be much lower than an LoA at which supposedly human free choices turn out to be entirely pre- and hetero-determined as well. Emagination is certainly nothing more than an emergent property of recording and preprogrammed configuration/reconfiguration. However, the same might be true to some extent of human imagination.

The thesis that will be advanced now is that one can distinguish among three levels, primarily temporally, but also 'in depth' of the relations between human and digital imagination. For Georges (2011, 40), digital identities are declarative, active, and calculated. The declarative element is made of all data voluntarily introduced by the user, notably during the process of inscription to the service—on Facebook, for instance, infor-mation about "Work and Education," "Contact and Basic Info" (birth-date, gender, religious view, et cetera), "Family and Relationships," et cetera. The active element consists of all the messages about the user's social activities—"x and y are now friends," "x and y shared z's photo," "x and y like z," and so on. The calculated element is represented by all the numbers resulting from the system's calculations, which are gener-ally visible on the user's profile—"you, x and y and n other like this"; "Friends n"; "view activity log n+"; "n more pending items"; "home n," and so on. But many of these numbers are, of course, invisible to the users, insofar as they represent the quantified elements used by the socio-technical systems for purposes such as control, surveillance, marketing, clustering, et cetera.

It is precisely this threefold distinction that I propose to transpose in the field of *emagination*. At the time of the Web 1.0, when the emplot-ment of our lifestories online was entirely dependent on the elements we voluntarily (if one excludes, of course, the several problems of addiction that such technologies could already provoke) entered into the virtual environment, the digital imagination then was 'above' human imagina-tion. With the rise of the social Web, the distance between the two imagi-nations has drastically reduced.

At the beginning of the 2000s, Manovich (2001, 218) used to oppose narrative and database. Narrative is linear, while database is a structured collection of data that can take different forms, hierarchical, network, relational, or object-oriented. New media, he said, and especially the Internet, are databases in a basic sense, since they "appear as collections of items on which the user can perform various operations: view, navigate, search" (219). However, today, databases are, at least from a user perspective, highly narrativized. As argued by Manovich (2012, np) himself more recently, "in social media, as it developed until now (2004–2012), database no longer rules. Instead, social media brings forward a new form: a data stream. Instead of browsing or searching a collection of objects, a user experiences the continuous flow of events."

Manovich (np) wrongly compares data streams to "a surrealist intentional juxtaposition of completely unrelated objects," and the data stream experience to that of the flaneur, who "navigates through the flows of passersby and the city streets, enjoying the density of stimuli and information provided by the modern metropolis." In this way, he implicitly readmits database as the main symbolic form of our culture. But in data streams, data is organized according to a temporal coherence that is meaningful for the user. For instance, this holds true of Facebook's News Feed or Twitter's timeline: "Twitter you are likely to care about most will show up first in your timeline. We choose them based on accounts you interact with most, Tweets you engage with, and much more." According to Nadav Hochman (2014. Emphasis mine), "if the database suppressed traditional linear forms (as it has no pre-defined notions of time), the data stream seems to emphasize once again the linearity of a particular data sequence. [. . .] Put differently, [. . .] *the stream brings back the temporal element as its core organizational and communicational factor.*"

Certainly, the temporality of data streams does not correspond to that of the classic narrative. Rather, David M. Barry (2013, 144) has said that one could think of the data stream as "distributed narratives which, although fragmentary, are running across and through multiple media, in a similar way to that Salman Rushdie evocatively described in *Haroun and the Sea of Stories.*" However, the data stream is the paradigm of a general tendency consisting of narrativizing the user experience on the Web, that is, in making it more familiar or even intimate for us. Let us consider, for instance, three features introduced in the last years by Facebook: Year in Review ("a collection of photos from your most significant moment this year"), On this Day (which "brings you memories to look back on from that particular day in your Facebook history"), the Friends Day Video (that "celebrates you and your friends"). In all these cases, algorithms operate on a database in order to give it a certain temporal and thematic coherence, that is, an emplotment. The logic of the digital and the logic of narrative, which is, as it has been shown, the logic of the productive imagination, draw close to each other.

Finally, in the age of Big Data and new algorithmic machines, it is as if *e*magination were breaking its link of familiarity with human imagination. The two imaginations are taking different paths, taking leave from each other, and, at least in terms of quantification and complexification, *e*magination is having the upper hand on human imagination. Jenna Burrel (2016, 1–2) has suggested that the most troubling opacity in machine learning algorithms is not opacity "as intentional corporate or institutional self-protection and concealment," nor opacity "stemming from the current state of affairs where writing (and reading) code is a specialist skill," but rather "an opacity that stems from the mismatch between mathematical optimization in high-dimensionality characteristic of machine learning and the demands of human-scale reasoning and styles of semantic interpretation."

The analogy between productive imagination as it has been presented in this chapter and the transcendental logic behind the process of data analytics as displayed by critical theorists Antoinette Rouvroy and Thomas Berns (2013) is astonishing. For them, data analytics is a threefold process: (1) the digitalization of life, which is the massive collection of data with purposes of security, control, optimization, marketing, et cetera; (2) data mining and machine learning, that is to say the algorithmic treatment of digital traces in order to extract significant correlations; and (3) the application of the extracted correlations, which is the profiling.[9] Similarly, Rob Kitchin (2014) presented data analytics as being made of (1) pre-analytics (data selection, data pre-processing, data reduction and projection, data enrichment); (2) machine learning, which consists of automatically learning to recognize complex patterns and construct models that explain and predict such patterns and optimize outcomes; and (3) data visualization and visual analytics, that is, diagrams, graphs, spatializations, maps, and animations that effectively reveal and communicate the structure, pattern, and trends of variables and their interconnections. Incidentally, it must be said that data visualization is just one possibility, the most 'anthropomorphic' of Big Data and new algorithmic machines' reconfigurations. Data visualizations are instruments that we need in order to orient ourselves in a dataworld that would otherwise remain in great part extraneous to us because of its vastness and complexity. But visualizations exist solely for our needs, and certainly not for that of the digital machines.

However, there are also at least three important differences between narrative imagination and Big Data analytics. First, in the latter, data are, at least in principle,[10] completely abstracted from their context of production. Second, data mining and machine learning are based neither on narrative emplotment nor on the research of causes (which, according to Aristotle, is the basis of scientific knowledge, and is what makes philosophy the first among sciences), but on the correlation of heterogeneous data. Third, in the case of data analytics, the application does not target a specific individual, but rather attributes the same behavioral predictions to all those who correspond to the same profile. For these reasons, one could say that Big Data and new algorithmic machines are sort of limit-cases of what I have called *e*magination, in which the digital operates through forms of 'sensibility' and 'understanding' which are radically different from those of human beings. As I will argue in the finale, familiarity and extraneity are the two sides of the current digital coin.

* * *

Rouvroy and Berns define algorithmic governmentality as "the automated extraction of pertinent information from massive databases for purposes of prevision or exclusion" (Rouvroy and Berns 2013, 167). According to Lucas D. Introna (2015, 27), the term stands for the performative nature of the new calculative practices: "In their flow of action, they [algorithms] enact objects of knowledge and subjects of practice in more or less significant ways. [. . .] Their actions are not just in the world, they make worlds." The fact is that productive imagination is not just a matter of cognition. In *Kant and the Problem of Metaphysics* (1997), Heidegger was conscious of the fact that Kant's theory of transcendental imagination is also a matter of human freedom.[11] The limit of the Kantian approach consists in having looked for freedom in the interiority of consciousness rather than in the thrownness of existence. In the course of this chapter, I have similarly argued that the schematism does not happen in the depth of the human soul, but rather outside, in the digital (but not only) environment that surrounds us. Yet according to the way imagination has been described, freedom does not consist in Heidegger's heroic and solitary authenticity. As argued by Verbeek (2011, 85) in his postphenomenological interpretation of Michel Foucault, "freedom does not consist in an absence of power but in gaining a new relation to power." Steve Dorrestijn (2012, 226) stressed that Foucault did not try to separate two spheres, the human and the technological (in the broad sense Foucault gave to the term "technology"), because "there is no genuine subject that is free from constraints or not yet affected by technology." Thus, freedom is less a matter of liberation than the possibility of recombining, and hence renegotiating, the frontiers with the sociotechnical environment that surrounds us. In other terms, human beings are essentially hetero-determinate, and what we call 'freedom' is a long and difficult detour through our technological, but also bodily, cultural, and social exteriorities. This implies, for instance, that there is no extension which is not at risk of becoming extraneous. More radically, this means that there is no extension which is not at risk of making us extraneous to ourselves. I am not arguing that extensions and exteriorizations, especially the technological and digital ones, are alienating per se. The fact is that human imagination cannot but be always-already externalized, because human beings are, as Helmuth Plessner put it, "artificial by nature" (De Mul 2014). I am just reaffirming the pharmacological principle that I have introduced in the previous part of this book. It is my belief that digital externalizations and extensions in the age of Big Data and new algorithmic machines have specific consequences on our subjectivities and sense of freedom. This will be the main argument of the finale. For the moment, I propose to take a step back, from the interferences between humans and non-humans to the symmetry between them. While in this chapter I have attributed emerging interpretational faculties to some digital machines, in the next I am going to wonder what imagination is in human beings and their contemporary cultural productions.

Notes

1. According to De Mul (2015, 107), customs, morals, and norms are no longer followed because no other possibilities are known, as in the pre-modern era. Neither are they chosen on the basis of deep conviction, as in modernity. Rather, "in this era, tradition has become a commodity rather that an existential choice. [. . .] [Traditions] are being 'shopped together' because they are useful or pleasant for the time being, but they are easily replaced by the 'next' tradition that makes its appearance in the media." I do not share his opinion according to which there is a radical difference between the 'narrative' cultures and identities of modernity, and the 'database' culture and identities of postmodernity. For me, as I am going to argue, algorithms are still emplotment or narrative machines, although of a specific kind. But I do share his idea that postmodern culture, imagination, and identities have less to do with engineering than with a form of bricolage. I will extensively discuss this point in the next chapter.
2. One can notice the similarities between the Ricoeurian triple *mimesis* and Callon, Lascoumes, and Barthe's idea of a threefold translation in the process of interaction between the scientific laboratory and the social reality in which it is immersed. Translation (1) has to do with a process of "reduction" of the macrocosm to the microcosm of the laboratory. Translation (2) is presented as an emplotment (although the authors do not use this term) of inscriptions—the world put into words. Finally, translation (3) aims at returning to the big world—"access to the world and to have a hold on it"—through the process of "laboratorization." The final result is Translation (with a capital *T*), which is "a partial reconfiguration of the macrocosm" (see Callon, Lascoumes, and Barthe 2009, 48–70).
3. For Ricoeur, metaphors are not entirely verbal. In metaphoric imagination there is also a pictorial and figurative dimension. This does not mean to resort to the old theory of imagination as the production of mental images of something absent. According to him, to imagine is not to have a mental picture of something but to display relationships in a depicting mode (Ricoeur 1978, 150).
4. This includes e-memory technologies. Incidentally, once autonomy, that is, the fact that these kinds of technologies do not only *store* data, but also *process* it, is considered to be one of the main features of e-memory technologies, we have already left the field of memory properly speaking and entered that of productive imagination.
5. In Clark, there is also a certain optimism or at least a lack of critical thinking about language. Language is for him an "artificial manipulation," a remedial to the connectionism that does not seem to allow abstract thinking. In his words, "[l]anguage is like that [. . .] because thought (or rather, biologically basic thought) is *not* like that" (Clark 2001, 47). In some sense,

language is the first, and probably also the most important, extension of human minds. And yet, he does not wonder about the fact that language is also probably the most powerful means through which social instances of domination and exclusion are embedded and reiterated. In other words, one can say that Clark's mind is extended, but not really socially and culturally situated.

6. Verbeek has 'expanded' these relations from below and from above. From below, he has introduced the notion of cyborg relations a radical variant of embodied relations in which "technologies actually *merge* with the human body" (Verbeek 2011, 144). This is the case, for instance, of psychopharmaca and neural implants. From above, he has spoken of immersion relations, in which technologies do not merge with the human body, but rather with the environment. Examples are smart environments from toilets and beds to entire cities (Rosenberger and Verbeek 2015, 21–22).

7. I am particularly thankful to Galit Wellner for this remark. In her words, "as an ex-programmer I can testify that coders try to neutralize software from data so that even min/max values are frequently put in an external file to avoid their coding into the system" (Wellner, personal communication). Wellner (2018) has recently described the posthuman, technologically mediated-hybridized, imagination as a multiple layering (consider the example of Instagram's filters or the logic of augmented reality). I see such perspective as complementary to mine.

8. https://affinitymap.epfl.ch/. Accessed June 10, 2019.

9. In this context, I do not expect to enter the algorithmic structure. Domingo (2015) has presented, for instance, five families of machines learning algorithms and schools, each one with its advantages and its limits: symbolist, connectionist, evolutionary, Bayesian, and analogizer—not to mention the fact that different approaches can be combined. If sporadically in this book I tend to privilege neural networks, it is for two reasons: (1) for some analogies they entertain with hermeneutics—see the conclusion of this part on this point; (2) because they are obtaining some great success in different applications—although an evolutionary algorithm has recently outperformed deep-learning machines at video games. See "Emerging Technology from the arXiv." www.technologyreview.com/s/611568/evolutionary-algorithm-outperforms-deep-learning-machines-at-video-games/. *MIT Technology Review*, July 18, 2018. Accessed June 10, 2019.

10. Latanya Sweeney (2000) found that combinations of few characteristics often combine in populations to uniquely or nearly uniquely identify some individuals. Hence, data released containing such information about these individuals must not be considered anonymous.

11. In Heidegger's book, one can find three main theses (Piercey 2011): (1) imagination is the common root of sensibility and understanding; (2) productive imagination in Kant is at the bottom of voluntary spontaneity and practical freedom; (3) Kant's ideas on imagination anticipate Heidegger's insights on the *Dasein*'s temporality and freedom, although without bringing them to the proper end. Kant refused to resort to productive imagination in his *Critique of Practical Reason*, which deals with moral philosophy. Yet, Johnson (1985, 265) argued that "in spite of his repeated insistence on the purely rational nature of moral judgment, Kant recognized the need for imagination in order to apply moral rules to specific cases." Furthermore, several contemporary thinkers such as Carol Gilligan, Martha Nussbaum, and especially Paul Ricoeur have insisted, mostly in the wave of Kant, on the importance of imagination for ethical and political actions (Pierron 2015).

References

Adorno, T., and Horkheimer, M. 2003. "The Culture Industry: Enlightenment as Mass Deception." In C. Jenks (ed.). *Culture: Critical Concepts in Sociology. Volume 2.* London and New York: Routledge, 146–180.

Amalric, J-L. 2014. *Paul Ricœur, l'imagination vive: Une genèse de la philosophie ricœurienne de l'imagination.* Paris: Hermann.

Barry, D. 2013. *The Philosophy of Software: Code and Mediation in the Digital Age.* London: Palgrave Macmillan.

Burrel, J. 2016. "How the Machine 'Thinks': Understanding Opacity in Machine Learning Algorithms." *Big Data & Society*, January–June, 1–12.

Callon, M., Lascoumes, P., and Barthe, Y. 2009. *Acting in an Uncertain World: An Essay on Technical Democracy.* Cambridge, MA: The MIT Press.

Castoriadis, C. 1994. "Radical Imagination and the Social Instituting Imaginary." In G. Robinson, and J. Rundell (eds.). *Rethinking Imagination: Culture and Creativity.* London and New York: Routledge, 136–154.

Castoriadis, C., and Ricoeur, P. 2016. *Dialogue sur l'histoire et l'imaginaire social.* Paris: Editions de l'EHESS.

Clark, A. 2011. *Supersizing the Mind: Embodiment, Action, and Cognitive Extension.* Oxford: Oxford University Press.

———. 2001. *Mindware: An Introduction to the Philosophy of Cognitive Science.* Oxford: Oxford University Press.

Clark, A., and Chalmers, D. 1998. "The Extended Mind." *Analysis* 58: 10–23.

Clowes, R. 2013. "The Cognitive Integration of E-Memory." *Review of Philosophy and Psychology* 4(1): 107–133.

De Mul, J. 2015. "Database Identity. Personal and Cultural Identity in the Age of Global Datafication." In W. de Been, P. Aurora, and M. Hildebrandt (eds.). *Crossroads in New Media, Identity and Law: The Shape of Diversity to Come.* London: Palgrave Macmillan, 97–118.

———. 2014. "Artificial by Nature: An Introduction to Plessner's Philosophical Anthropology." In J. De Mul (ed.). *Plessner's Philosophical Anthropology: Perspectives and Prospects.* Amsterdam: Amsterdam University Press, 11–37.

Domingo, Pedro. 2015. *The Master Algorithm: How the Quest for the Ultimate Learning Machine Will remake Our World.* New York: Basic Books.

Dorrestijn, S. 2012. "Technical Mediation and Subjectivation: Tracing and Extending Foucault's Philosophy of Technology." *Philosophy & Technology* 25(2): 221–241.

Floridi, L., and Sanders, J. 2004. "On the Morality of Artificial Agents." *Mind & Machines* 14(3): 349–379.

Foessel, M. 2007. "Paul Ricœur ou les puissances de l'imaginaire." In M. Foessel, and F. Lamouche (eds.). *Paul Ricœur, Anthologie.* Paris: Seuil, 7–22.

Georges, Fanny. 2011. "L'identité numérique sous emprise culturelle: De l'expression de soi à sa standardisation." *Les cahiers du numérique* 7(1): 31–48.

Heidegger, M. 1997. *Kant and the Problem of Metaphysics.* Bloomington: Indiana University Press.

Hill, R. 2016. "What an Algorithm Is." *Philosophy & Technology* 29(1): 35–59.

Hochman, N. 2014. "The Social Media Image." *Big Data & Society*, August. http://bds.sagepub.com/content/1/2/2053951714546645. Accessed June 10, 2019.

Hui, Y. 2016. *On the Existence of Digital Objects*. Minneapolis: Minnesota University Press.

Ihde, D. 1990. *Technology and the Lifeworld: From Garden to Earth*. Bloomington: Indiana University Press.

Introna, L. 2015. "Algorithms, Governance, and Governmentality. On Governing Academic Writing." *Science, Technology & Human Values* 41(1): 17–49.

Johnson, M. 1985. "Imagination in Moral Judgment." *Philosophy and Phenomenology Research* 46(2): 265–280.

Kant, I. 1998. *Critique of Pure Reason*. Cambridge: Cambridge University Press.

Kearney, R. 1998. *Poetics of Imagining: Modern to Post-Modern. 2nd edition*. New York: Fordham University Press.

Kitchin, R. 2014. *The Data Revolution: Big Data, Open Data, Data Infrastructures and Their Consequences*. London: Sage.

Lenk, H. 1995. *Schemaspiele: Über Schemainterpretationen und Interpretationskonstrukte*. Frankfurt am Main: Suhrkamp.

Makkreel, R. 1990. *Imagination and Interpretation in Kant: The Hermeneutical Import of the Critique of Judgement*. Chicago: The University of Chicago Press.

Manovich, L. 2013a. *Software Takes Command*. London: Bloomsbury.

———. 2013b. "Software Is the Message." http://lab.softwarestudies.com/2013/12/software-is-message-new-mini-article.html. Accessed June 10, 2019.

———. 2012. "Data Stream, Database, Timeline." http://lab.softwarestudies.com/2012/10/data-stream-database-timeline-new.html. Accessed June 10, 2019.

———. 2001. *The Language of New Media*. Cambridge, MA: The MIT Press.

Mittelstadt, B.D., Allo, P., Taddeo, M., Wachter, S., and Floridi, L. 2016. "The Ethics of Algorithms: Mapping the Debate." *Big Data & Society*, July–September, 1–21.

Noë, A. 2009. *Out of Our Heads: Why You Are Not Your Brain, and Other Lessons from the Biology of Consciousness*. New York: Hill and Wang.

Piercey, R. 2011. "Kant and the Problem of Hermeneutics: Heidegger and Ricoeur on thre Transcendental Schematism." *Idealistic Studies* 41(3): 187–202.

Pierron, J-P. 2015. "Imaginer plus pour agir mieux: L'imagination en morale chez Carol Gilligan, Martha Nussbaum et Paul Ricœur." *Les ateliers de l'éthique/The Ethics Forum* 10(3): 101–121.

Ricoeur, P. 2016. "Architecture and Narrativity." *Études Ricœuriennes/Ricoeur Studies* 7(2): 31–42.

———. 2004. *The Rule of Metaphor*. New York and London: Routledge.

———. 2002. *Cinque lezioni: dal linguaggio all'immaginae*. Palermo: Aesthetica Print.

———. 1992. *Oneself as Another*. Chicago: The University of Chicago Press.

———. 1991. *From Text to Action: Essays in Hermeneutics, II*. Evanston: Northwestern University Press.

———. 1988. *Time and Narrative, Volume 3*. Chicago: The University of Chicago Press.

———. 1986. *Time and Narrative, Volume 2*. Chicago: The University of Chicago Press.

———. 1978. "The Metaphorical Process as Cognition, Imagination, and Feeling." *Critical Inquiry* 5(1): 143–159.

Rodighiero, D., and Romele, A. Forthcoming. "The Hermeneutic Circle of Data Visualization: The Case of the Affinity Map." *Techné: Research in Philosophy and Technology*.

Romele, A., and Severo, M. 2016. "From Philosopher to Network: Using Digital Traces for Understanding Paul Ricoeur's Legacy." *Azimuth* 7: 113–128.

Rosenberger, R., and Verbeek, P-P. 2015. "A Field Guide to Postphenomenology." In R. Rosenberger, and P-P. Verbeek (eds.). *Postphenomenological Investigations: Essays on Human-Technology Relations.* London: Lexington Books, 9–42.

Rouvroy, A., and Berns, T. 2013. "Gouvernementalité algorithmique et perspectives d'émancipation." *Réseaux* 1(177): 163–196.

Smolicki, J. 2018. "You Press the Button, We Do the Rest. Personal Archiving in Capture Culture." In A. Romele, and E. Terrone (eds.). *Towards a Philosophy of Digital Media.* London: Palgrave, 77–100.

Stiegler, B. 2014. "Pharmacologie de l'épistèmè numérique." In B. Stiegler (ed.). *Digital Studies: organologie des savoirs et technologies de la connaissance.* Limoges: FYP Éditions, 13–26.

———. 2001. *Technics and Time 3: The Cinematic Time and the Question of Malaise.* Stanford: Stanford University Press.

Sweeney, L. 2000. "Simple Demographics Often Identify People Uniquely." *Carnegie Mellon University, Data Privacy Working Paper 3.* http://dataprivacylab.org/projects/identifiability/paper1.pdf. Accessed June 10, 2019.

Taylor, G.H. 2006. "Ricoeur's Philosophy of Imagination." *Journal of French Philosophy* 16(1–2): 93–104.

Thompson, E. 2007. *Mind in Life: Biology, Phenomenology, and the Sciences of Mind.* Cambridge, MA: Harvard University Press.

van den Boomen, M. 2014. *Transcoding the Digital. How Metaphors Matter in New Media.* Amsterdam: Institute of Network Cultures.

Van Den Eede, Y. 2015. "Tracing the Tracker. A Postphenomenological Inquiry into Self-Tracking Technologies." In R. Rosenberg, and P-P. Verbeek (eds.). *Postphenomenological Investigations: Essays on Human-Technology Relations.* London: Lexington Books, 143–158.

———. 2010. "In Between Us: On the Transparency and Opacity of Technological Mediation." *Foundations of Science* 16(2): 139–159.

Verbeek, P.P. 2011. *Moralizing Technology: Understanding and Designing the Morality of Things.* Chicago: The University of Chicago Press.

Wellner, G. 2018. "Posthuman Imagination: From Modernity to Augmented Reality." *Journal of Posthuman Studies* 2(1): 45–66.

Wiener, N. 1964. *God & Golem, Inc. A Comment on Certain Points Where Cybernetics Impinges on Religion.* Cambridge, MA: The MIT Press.

4 We Have Never Been Engineers

In the previous chapter, I proposed a sort of exaltation of new digital machines by attributing to some of them an emergent imaginative capacity. In this chapter, I want to take an opposite route, which consists in humiliating human claims in terms of imagination and creativity. If digital imagination resembles human imagination, it is not just because nowadays numerous machines are capable of amazing performances. This also happens because human beings turn out to be more ordinary than they are willing to admit. I contend that 'we have never been engineers' and that at most we can define ourselves as *bricoleurs*.

I will further deal with the distinction between engineers and *bricoleurs*, first introduced by Claude Lévi-Strauss. In these preliminary remarks, I would rather like to focus on the similar distinction Michel de Certeau has suggested between "strategy" and "tactic." De Certeau himself repeatedly called upon the concept of bricolage as an analogous of his notion of tactic. This is the case, for example, when he talks about reading. Against the dominant idea that reading is a passive practice, he puts forward the idea (strongly hermeneutic, one could say) according to which text is a construction of the reader, who gives rise to something different from the intention of the text and its author. This activity, writes de Certeau (1984, 174), can be considered "as a form of the bricolage Lévi-Strauss analyzes as a feature of the 'savage mind,' that is, an arrangement made with 'the materials at hand,' a production 'that has no relationship to a project,' and which readjusts 'the residues of previous construction and destruction.' "

The prime intent of de Certeau is to bring forward the consumers' ways of production. By massively consuming products of the cultural industry such as texts and images, we also end up very often reconfiguring them differently. The same holds true for other cases such as urban spaces, foods and recipes, et cetera. This means that there is a manipulation by users that are not makers that deserve to be analyzed. It is not a minor phenomenon concerning small groups such as the primary societies that anthropologists and ethnographers study, but something having to do with a silent majority. After all, who is not a consumer today in some respects? If someone has the privilege of being a producer in certain domains, she will certainly be a consumer in others. In the words of de Certeau (XVII),

> Marginality is today no longer limited to minority groups, but is rather massive and pervasive; this cultural activity of the non-producers of culture [. . .] remains the only one possible for all those who nevertheless buy and pay for the showy products through which a productivist economy articulates itself. Marginality is becoming universal. A marginal group has become now a silent majority.

And it is precisely the consumer that we all are, to some extent, who is oriented toward tactics rather than strategy. Strategy has to do with "the calculation (or manipulation) of power relationships that becomes possible as soon as a subject with will and power (a business, an army, a city, a scientific institution) can be isolated" (35–36). Central to such a notion are the concepts of "place," "proper," and "exteriority": "It [strategy] postulates a *place* that can be delimited as its *own* and serve as the base from which relations with an *exteriority* composed of targets or threats [. . .] can be managed" (36). One could say that the strategy is related to the possibility of exercising an internal absolute control (de Certeau speaks of "*panoptic practice*") on a place which has been previously separated and isolated from its surroundings, in order to defend itself and eventually to counterattack. The logic behind the strategy is not much different from that of Carl Schmitt's total State—and the distinction between "inside" and "outside" is similar to the Schmittian one between "friend" and "enemy."

The tactic is a calculated action as well, but the difference lies in the fact that it is characterized by the absence of the "proper": "a tactic is a calculated action determined by the absence of a proper locus. No delimitation of an exteriority, then, provides it with the condition necessary for autonomy. The space of a tactic is the space of the other. [. . .] It [the tactic] can be where it is least expected. It is a guileful ruse. In short, a tactic is an art of the weak" (36–37). In sum, tactic regards the capacity and the possibility of making do and moving within a place that has not been seized according to our will or need. It has to do with the room of maneuvering that exists within an already-codified place in which we arrive, so to say, late. Those who develop tactics might be seen as the internal enemies of Schmitt—and it is not by chance that de Certeau relies on Clausewitz in the passages about tactic.

The thesis I would like to develop is that the distinction between strategy and tactic is mostly a false distinction, and that in reality we all cannot but resort to the latter. This does not mean, of course, that there are not different degrees in being active or passive, auto- or hetero-determined in our relationships with the world. But for me it is important to stress that the idea of strategy falls in great part under a sort of demiurgic myth. All governments, for instance, are concerned with the choices that have been previously taken, with the global economy, industry, et cetera. There is no individual that does not have to deal with a language they have inherited, the rules of a field that they want to occupy, et cetera. Such a general rule of hetero-determination is valid at a social, historical, cognitive, bodily, and technological level, at least. And this holds true even for the most powerful (the Leviathan who has created the greatest number of alliances and translations, in the terminology of Callon and Latour) actor, for big tech companies like Google and Facebook. For this reason, I cannot share de Certeau's tendency to *essentially* distinguish between two ways of action, one of the weak and one of the powerful. Tactic is for me the only way in which human freedom, and more generally our dealing with the world, manifests itself at both an individual and collective level.

Let's come back to the specific issue of imagination, which will be at the center of my interests in this chapter as well. In a recent article entitled "Against Creativity" (2018), Alison Hills and Alexander Bird affirm that imagination has not to do with absolute novelty. A scientist, for instance, "chooses her problems in large part on the basis of their similarity or analogy to existing problems. And she reaches for possible solutions by considering analogues of solutions to existing problems" (5). The analogical thinking is at the core of the scientific imagination. The same holds true for artistic production:

> Beethoven closely followed the models of his predecessors, especially Haydn, but also Mozart. So, for example, Beethoven uses exactly the same orchestra as Haydn; the symphony has the same number and kind of movements as its predecessors; the first movement follows classical sonata form to the letter, and like several of Haydn's later symphonies it opens with a slow introduction before the first subject appears. Analogies appear at the level of detail also, with many ideas resembling those found elsewhere: basing the first subject of the first movement around an ascending arpeggio is a very common device; in several places there are strong echoes of ideas in his predecessors' symphonies (e.g. Mozart's *Jupiter* symphony). Adopting a certain model also poses problems to be solved: the second subject would be expected to contrast with the first, so Beethoven uses a descending scale; the slow introduction cannot be unrelated to the rather fast main body of the first movement, so Beethoven has to deploy ideas in the latter that recall the former, and so on. The creative process evolves, subject to a considerable variety of constraints.
>
> (Ibid.)

To be imaginative does not mean to randomly move in a vacuum, but rather to learn to orient oneself within a delimited space according to a certain number of rules. The more the limits and the rules are known, the more one can 'freely' move among them. And supposedly one can also modify or at least suggest some modifications of them. So both in art and in science it is not a matter of limiting oneself to the imitation of what has already been done—although imitation as well as reproduction have an intrinsic value, both in art and science, which is often neglected (let's think, for instance, of reproducibility). It is rather a matter of reconfiguring—that is, synthetizing, narrativizing or schematizing differently—the already-existing elements, and eventually of adding a few new blocks of knowledge, practice or products. Creativity, imagination, and novelty are not much more than a rest, a deviation or, as I will argue below, a shibboleth: "Of course, unless her purpose is to create reproductions, she [the artist] must not imitate them [valuable works of art that already exist] exactly. Nor should she attempt to extract rules of artistic creation. Rather she will treat these works, and the tradition from which they come, as guides and exemplars" (17).

The following chapter is developed in three sections. In the first section (4.1), I will further argue in favor of the centrality of the notion of imagination in the field of new media and technologies studies. In particular, I will resort to the thought of the mostly unknown French thinker Robert Estival, who between 2001 and 2003 published the three volumes of his *General Theory of Schematization*. I will also speak a few words on the schematization as theorized by Gilbert Simondon. In the second section (4.2), I am going to introduce a distinction between imagination-engineer and imagination-bricolage, and I will argue against the existence of the former. In the third section (4.3), I will return from human beings to digital machines, and I am going to present some reflections on the recent forms of automatic art.

4.1 The General Theory of Schematization

While in the first part of this book I have mostly undertaken a series of reflections on the primacy of registration, recording, and keeping track over information and communication to understand digital media and technologies as they are today, this second part is mainly devoted to the emergence of a third paradigm I have previously called *emagination*. For me, emagination does not contradict registration. It can rather be seen as an emerging property of it, just as registration can be seen as an emerging property of information and communication.

In this section, I would like to further discuss the concept of emagination by using a bizarre pair of glasses: the thought of a mostly unknown French thinker, Robert Estivals, who wrote a *General Theory of Schematization* (*Théorie générale de la schematisation*, GTS), published in three volumes between 2001 and 2003. Estivals has been professor in communication and media studies at the University of Bordeaux

Montaigne between 1968 and 1993. Incidentally, he was not just an academic, but also an artist; he founded several avant-garde movements. In 1996 he created a museum in his hometown, Noyers-sur-Serein, called La Maison du Schématisme. Two aspects are particularly interesting of his approach.

First, Estivals distinguishes among three phases in the field of communication studies or, to be more precise, among three phases "in the evolution of the canonical schemas of communication." The first one, between Saussure and the 1960s, used to resort to an interpersonal schema, recognizing in this way the specific field of "communicology": the relation between entities. Let us think back, for instance, to the famous schema in Saussure's *Course in General Linguistic*, in which two persons discuss, and something seems to happen in their minds; and to its geometric version, in which it is clear that what is happening in these minds, in the coming and going of audition and phonation, is the relation between the acoustic image and the concept. According to Estivals (2003, 46), the archetypes of this schema are the line and the feedback. The canonical schema of communication is both linear and circular, as far as it privileges the enunciation, on the one hand, but on the other hand it also implies the possibility for the receiver to retake the initiative. Another schema of this kind is Shannon and Weaver's model of communication in which new concepts are introduced: information source, message, transmitter, signal, and so on.

In the second phase, between the 1960s and the 1990s, the debate shifted from interpersonal to social communication. The archetypes of the line and the feedback gave way to that of the network. Among others, Estivals proposes a schema by the electrical engineer Abraham Moles, who was also professor of sociology and social psychology at the University of Strasbourg, and is today considered one of the pioneers of communication studies in France. In the third and last phase, that began in the 1990s, the concern of communication studies for linguistic, information engineering and sociology declined, and cognitive sciences took the upper hand: neurology, psychology, logic, and especially artificial intelligence. The archetype, in this case, is Franck Rosenblatt's perceptron, created in 1958, that Estivals (64) also calls "networked machines."

This periodization has strong analogies with the one that characterizes the more recent history of digital media, and which was presented in the previous chapter through the adaptation of the three dimensions of digital identity according to the French media theorist Fanny Georges. Moreover, the shift from social networks to neural networks Estivals speaks about is very close to the transition from recording and registration to *emagination* I am going to debate. In the first part of the book, I have argued that the classic information theory cannot account for the digital and its current impact on society. What is at stake today is not so much information circulation, but the fact that all flow of information, and more generally all digitally mediated or accompanied action, leaves a voluntary, involuntary, or induced digital trace.

Here, my thesis is that we are entering a third phase, in which not just information and registration are at stake, but also emagination. And this is the second reason that has made me appreciate the works of Robert Estivals. In the GTS, he argues that communication has to do with schematization, and this is, as I have shown, the specific cognitive task Kant attributed to imagination in the first *Critique*. In the first volume of the GTS, the author explicitly refers to Kant, affirming that "schematism and scheme are then considered [by Kant] as the intermediary mechanism allowing [. . .] to go from the plurality to the unity. Kant does not deal with the concept of information's reduction, although such concept is implicitly treated" (Estivals 2001, 55). Other important references for him are the French psychiatrist Gabriel Revault d'Allones, Bergson, Flach, Sartre, Piaget, and the new generation of researchers of the years 1985–1995 that made the link between scheme and cognition, neurosciences, logic, and language: Andler, Smolensky, and Woodfield. In particular, Estivals considers Revault d'Allones as the true theoretician of schematism: firstly, because he proposed an encompassing and mostly inductive, theory of knowledge based on schematism, distinguishing among different steps; secondly, because not only did he not reduce schematism to an information's selection, but also to its possible organization. Revault d'Allones himself understood his theory as a generalization of Kantian schematism: "Extended as such, the function of schematisation, intermediary as Kant says between 'sensibility' and 'understanding,' [. . .] embraces the totality of the psychic life, and we can say the same of habits, memory, and even of association and attention. [. . .] It [schematization] presides over the knowledge of things and of ourselves, it has sensible forms, it has active forms, staggered from the psychic reflex to the deliberate will" (in Estivals 2001, 58–59).

Estivals (2003, 80) defines schematization as the "whole of the processes of cognition and symbolization based on two complementary principles: the reduction and the organization of information, the arborescent and the reticular schematization." He also affirms that "everything happens as if the first act of human understanding is aimed at the reduction of information [. . .]. Understanding is first of all reducing and isolating. [. . .] However, once the unity is achieved, the thought is led to involve concepts in relation" (81). The example he proposes is that of working with a computer: looking for and selecting information, on the one hand, and constituting a corpus, a hypertext, or a thesaurus, on the other hand. He also distinguishes between scheme, the operation of schematization in the human minds, and schema, which is its external manifestation, and can be linguistic, phrasal, textual, narrative, or metatextual. For him, communication is a series of relations of schematization, the scheme in the sender, the schema, which is its externalization, and the schema in the receiver. To tell the truth, I am oversimplifying Estivals's theory slightly: on the basis of Piaget's work, he distinguishes, indeed, in every individual, between an intuition scheme and a scheme of the cognitive structure. Moreover, at the level of the schema, he introduces a series of passages and distinctions between linguistic schema, transcoding, graphic schema, gestural schema, and so on. In any case, I consider his approach interesting, because I also believe that communication and interpretation are forms of schematization, that is of simplification and organization of recorded information about reality.

The limits of the GTS are, however, clear. First, there is an unjustified priority that is attributed to the linguistic schemas, while it has been noticed, for instance, that also gestures can be considered as tools for the synthesis that characterizes schematism and hence imagination (Maddalena 2015). Secondly, and more importantly, can we really say that schemas are mere externalizations of internal schemes? In my opinion, it is rather the opposite, in the sense that schemes can be seen as internalizations of the external schemas; or, to be less extreme, our schematizations often depend, with different possible degrees, on those syntheses that happen "out of our heads."[1] The literature that helps me in supporting my thesis is vast and comes from different schools of thought and approaches. I have already discussed Clark and Chalmers's "extended mind," as well as Stiegler's "tertiary retention." One could also mention Jack Goody's "graphic reason," adapted to computer sciences by Bruno Bachimont (2010), who introduced the notion of "computational reason." In this context, I would like to focus, for a moment, on the perspective of Gilbert Simondon, who specifically devoted a seminar to imagination and invention in 1965–1966 (Simondon 2014).

Simondon is one of the rare French thinkers who did not follow the linguistic turn in those years. He understands imagination as both externalized and technicized. According to Vincent Beaubois (2015, np), Simondon's imagination can be defined as practical and technical schematism. Indeed, the scheme is not a mental entity for him. It is rather an operation that is made in and by the things themselves. Our imagination consists in the capacity of perceiving some qualities in the things that are neither directly sensory nor are entirely geometrical. This is precisely the intermediary level of the schemes (here, of course, there is no distinction anymore between scheme and schema), which are in between matter and forms. In other words, human imagination is the intuition of the schematization that happens in the things themselves. For human beings, invention is the possibility of establishing new analogies among different technological lineages. There is, then, not just passive intuition but also a procedural activity in human imagination. Such an activity, Beaubois points out, is not merely conceptual, but directly depends on the practical and even emotional frequentation of the technical objects. What is interesting for the rest of this chapter is also the fact that if invention is the capacity of finding out formerly unseen analogies in the technical world already at our disposal, then imagination has less to do with pure invention than with recombination. According to Simondon, the inventor does not proceed ex nihilo but from already technical elements. Similarly, in the "circle of the images" he proposes at the beginning of the 1965–1966 seminar, invention itself is presented not as pure creation, but as reorganization of the system of images (the term "image" stands often in this seminar for "scheme") which is already in our possession.

4.2 Engineers vs *Bricoleurs*

The case of digital technologies and media, or at least an emergent part of them, is particular. They do not just support or mediate our schematizations, but also they schematize by themselves. As De Mul (2018) points out, whereas in old technologies such as writing the products of thinking are outsourced to an external memory, in the case of digital technologies thinking itself is outsourced to an external device.

However, before going back to emagination, it remains to be seen what I mean by imagination, tout court. The history of philosophy has been marked by an alternative between two notions of imagination, which is somehow exemplified by the tension in Kant between imagination as described in the first and in the third *Critique*, respectively. In the *Critique of Pure Reason*, productive imagination appears as the mediating link between the passivity of impressions and the activity of the judgment of concepts. It is, then, a cognitive function that simplifies and articulates sense data, giving them a unity, and hence a meaning. It is schematization. In the *Critique of Judgment*, instead, imagination does not schematize (or at least it does not schematize properly; that is, according to the concept. In fact, in the *Critique of Aesthetic Judgment*, which is the first part of the third *Critique*, Kant seems still to be aware that imagination is always an interplay between spontaneity and the capacity of giving form). There is a "free play" between imagination and understanding. And imagination definitely frees itself from the concept in the aesthetic ideas, to which "no [determinate] *concept* can be adequate, so that no language can express completely and allow us to grasp it" (Kant 1987, 182). In the second part of the third *Critique*, the *Analytic of the Sublime*, and more precisely in § 49, the German philosopher speaks of an imagination that is not productive anymore, but creative, free from all law of association and nature. It is the imagination of the artist and the genius: the painter, the sculptor, and the poet.

In this context, I cannot debate the inconsistency of the clear-cut distinction, so classic in philosophy (let us think back to Bachelard), between the kinds of imagination at work in technoscience and poetry. I am restricting myself by referring the reader back to the already cited work of Max Black (1962) on models and metaphors, or to the "puzzling cases" of semi-arts and semi-techniques such as industrial design, architecture, and data visualization. According to Vilém Flusser (1999, 18), for instance, the sharp division between the world of the arts and that of technology and machines is a result of the modern bourgeois culture: "hence culture was split into two mutually exclusive branches: one scientific, quantifiable and 'hard,' the other aesthetic, evaluative and 'soft.' [. . .] In the gap, the word *design* formed a bridge between the two." On this basis, I propose to call the two notions (productive and creative) of imagination I have just presented through Kant, "imagination-bricolage" and "imagination-engineering," respectively. I am clearly referring to the distinction between *bricoleur* and engineer introduced by Claude Lévi-Strauss in *The Savage Mind* (1966).

For the French anthropologist, the starting point is the "Neolithic Paradox"—that is, the long stagnation of human scientific and technological development after the great advancements of that time: pottery, weaving, agriculture, domestication of animals, and so on. According to him (15), "there is only one solution to the paradox, namely, that there are two distinct modes of scientific thought [. . .] one roughly adapted to the stage of perception and the imagination; the other at a remove from it [. . .] one very close to, the other more remote from, sensible intuition." It is precisely in order to find an analogy today and here, in Western societies, of the former kind of scientific thought, closer to perception and imagination, not "primitive" but "prior," that he introduces the notion of bricolage. "The *bricoleur*," Lévi-Strauss (17) explains, "is adept at performing a large number of diverse tasks; but, unlike the engineer, he does not subordinate each of them to the availability of raw materials and tools conceived and procured for the purpose of the project." Rather, he goes on, "his universe of instruments is closed and the rules of his game are always to make do with 'whatever is at hand'"; the expression in French is "*moyens du bord*," the literal meaning of which is "the means at [disposal on] board," on a boat, for instance, when the land is still far away, and one has to make do. These resources represent, he adds in the same passage, "a set of tools and materials which is always finite and is also heterogeneous because what it contains [. . .] is the contingent result of all the occasions there have been to renew and enrich the stock."

When I first read this description, I immediately thought back to my father's garage, where there is no room for a car anymore, but one can find an old motor scooter, a collection of graphic novels, tools for gardening, for home maintenance and repairing, for cutting wood and metals, old shoes, new shoes for hiking, and many other things that seem to offer a solution to every domestic problem and need. My father is a *bricoleur*; I am not. And yet, as I will argue in the following lines, we have the same spirit, and the same imagination at work. For Lévi-Strauss (19), the difference between the engineer and the *bricoleur* is clear, because while the former "is always trying to make his way out of and go beyond the constraints imposed by a particular state of civilization," the latter "by inclination or necessity always remains within them." The engineer uses the concept, while the *bricoleur* resorts to the signs, whose specificity consists in being both material and conceptual, requiring then "the interposing and incorporation of a certain amount of human culture" (20). He also affirms, a few sentences below, that the concepts are operators "opening up," while signs and signification have to do with the mere "reorganization."

It is precisely this distinction, the existence of an engineer capable of working with many concepts and practically without words, on the one hand, and a *bricoleur* without much inventiveness, on the other hand, that has been generally problematized and, in some cases, sharply criticized (Mélice 2009). For instance, Derrida (1978, 285) says that "the engineer, whom Lévi-Strauss opposes to the *bricoleur*, should be the one to construct the totality of his language, syntax, and lexicon. In this sense the engineer is a myth." And he continues by affirming that "a subject who supposedly would construct it 'out of nothing,' 'out of whole cloth,' would be the creator of the verb, the verb itself. The notion of the engineer who supposedly breaks with all forms of bricolage is therefore a theological idea." Which means, in other words, that we as human beings all are *bricoleurs*, and that it has never been and never will be otherwise. We have never been engineers.

This same statement holds true for imagination. The imagination-engineering is a myth. Like my father, I am a *bricoleur*, as far as my knowledge, my writings, and even my most brilliant intuitions are concerned. They all depend on my competences in history of philosophy and its theory. I just *bricole* with words and thoughts, rather than with things and tools. Novelty and creation are no more than a shibboleth, a minimal deviation from one recombination to another. On this point, Maurizio Ferraris (2012, 316) points out: "Now, it is this very error, this shibboleth, that picks out the individuality of individuals and characterizes their uniqueness." For him, there are two ways of explaining the uniqueness of persons. The first one is "hifalutin and solemn, and it insists on our positive exceptionalness because we are infinite and ineffable," as when one talks of geniality and creativity. The second one, and this is the way I see imagination, defines uniqueness as "a negative exceptionality, a production error so to say."

In an article entitled "Structure and Hermeneutics," originally published in 1963 (one year after the publication of *The Savage Mind*), Ricoeur criticizes Lévi-Strauss for having indiscriminately applied structuralism, which works for understanding cold (totemic and synchronic) societies (that is, societies that pretend to ignore history), to the hot (kerygmatic and diachronic) ones (that is, to the societies in which the event, and hence the time of history, is always welcomed). He also explicitly refuses the concept of bricolage for our biblical and Western (and white?)

societies, which, according to him, always make room for the event: "*bri-colage* works with debris; [. . .] the debris plays the roles of a precon-straint. [. . .] [T]he reuse of biblical symbols in our cultural domains rests, on the contrary, on a semantic richness, on a surplus of what is signified, which opens toward new interpretations" (Ricoeur 2004, 45). Moreo-ver, if kerygmatic societies can be in minimal part understood through structuralism (Ricoeur speaks of a "secondary level of expression, sub-ordinated to the surplus of meaning found in the symbolic substratum"), totemic societies are not totally immune to the event either: "A good example to consider is the homology between marriage rules and food prohibition [. . .]; the analogy between eating and marrying, between fasting and chastity, constitutes a metaphorical relation anterior to the [structuralist] operation of transformation" (53). I do not want to enter, in this context, the debate about opportunity, from the point of view of cross- and intercultural respect, of *essentially* distinguishing between two forms of societies. Indeed, this can easily end up as a form of discrimina-tion or exclusion. I limit myself to stress that (1) over the years, Ricoeur will tend to change his mind about symbols. He will move toward Lévi-Strauss, while continuing to insist on the possibility of a minimal event (a sort of deviation) in the interstices of the structure. The imagination-bricolage I am presenting in this paragraph suggests something very simi-lar; (2) the representation of kerygmatic societies given by Ricoeur more than fifty years ago is far from our multicultural experiences and societies.

It is, I believe, the passage from exteriority, the passage we have seen in Simondon, which has been theorized by other philosophers but neglected by a long tradition, that brings us to temper our expectations toward human imagination and schematism. Such a passage, through language, culture, society, technology, body, and nature, grounds imagination in the world, namely in the limits and possibilities (the affordances) offered to us by the reality itself.

4.3 Automatic Art

And yet, we *see* novelty, we *see* creativity, and we *see* originality, especially in fields such as the art world. Interestingly enough, Lévi-Strauss collo-cates the artist halfway between the engineer and the *bricoleur*, because "by his craftsmanship he constructs a material object which is also an object of knowledge" (Lévi-Strauss 1966, 22). Which means that the art-ist is the one who does or provokes thoughts via things: colors, brushes, canvas, and so on. However, this is still a simplistic way to present the bricolage of the artist. There is, according to me, a deeper way that has to do with the very essence of the art world. Let us think of an artwork, a sculpture representing the US president Donald Trump struck down by a meteorite on a red carpet, and shards of glass next to his body. And sup-pose two visitors are at the gallery where this piece is exhibited, Alice and

Arthur. Alice does not know much about art, but she is very interested and engaged in politics. She is particularly horrified about Trump's idea of building a wall along the US-Mexico border, one of the famous central arguments of his presidential campaign. She also read in *Vanity Fair* that President Trump would like this wall to be "transparent," because drug dealers may otherwise throw large bags of drugs across to the other side, and hit innocent passers-by.[2] When she sees the wax sculpture in the gallery, the meteorite, and the pieces of glass, she is immediately reminded of the "transparent" wall, and of the sixty-pound bags thrown by drug dealers. She smiles, she thinks this artwork is provocative, she sees political criticism and satire in it. She identifies with the intentions of the author. In sum, she loves it.

Arthur is not much interested in politics. Of course, he knows about the wall, and he has heard from his friend Alice about the idea of a "transparent" wall, but he is rather resigned to the stupidity of human beings as political animals. For this reason, for some years now, he has sought refuge in the study of art history and theory. He also loves to frequent art galleries and exhibitions. In 2017, he went to Paris, and he visited the Italian artist Maurizio Cattelan's exhibition, *Not Afraid of Love*, at the Monnaie de Paris. Arthur had not been enthusiastic about art in 2011, when Cattelan had his retrospective at the Guggenheim in New York. In Paris, he finally saw the much-discussed installation *La nona ora* (The ninth hour), a life-sized effigy of Pope John Paul II hit by a meteor, on a red carpet, with pieces of glass next to him, that were first exhibited in 1999 at the Kunsthalle Basel. He loved that work, its title, and the multiplicity of its possible interpretations. And of course, he did not appreciate the Trump statue, which is for him nothing but a bad imitation of Cattelan's installation. At the very least, the artist could have replaced the meteorite with a large bag, and the red carpet with the dry field that characterizes many parts of the US-Mexico border. Contemporary art and culture is essentially quotationist, but one has to be careful about the fine line between quotation and plagiarism.

With this example I want to show how the appreciation of an artwork depends on the competence we have concerning the art world, which is essentially made of art history and theory, and to a certain degree of society: institutions, critics, galleries, and so on. At least, this is Arthur Danto's thesis in his major work, *The Transfiguration of the Commonplace* (1981). For him, there is an intrinsic difference between aesthetics and the philosophy of art. Indeed, aesthetics remains related to the *aisthesis*, which is the Greek word for "perception." But art, or at least art from the second half of the twentieth century on, has abandoned the realm of sensibility and landed into the field of pure intelligibility. Incidentally, this is also the reason why Danto will share Hegel's idea about the death of art, which has been henceforth absorbed by philosophy. The book starts, for instance, with the example of a series of paintings, all red rectangles, one next to the other. Although identical from an aesthetical (that is, perceptive) point of view, they are very different from an artistic point of view: they all represent different ideas, intentions, and artistic currents. In the case of the last rectangle described by Danto, a canvas grounded in red lead, we are actually not dealing with art at all. The fact is that there is no intrinsic and physical/material element that allow us to discriminate among them. Another famous example by Danto (120–121) is that of two paintings commissioned by a library of science to be executed on facing walls. Two artists are commissioned, J and his archrival K, and because of this rivalry their works are carried out with the greatest secrecy. When the veils fall on the day of inauguration, the two paintings look exactly the same: two vertical rectangles crossed by a straight line in their middle. The two works aesthetically *appear* as the same, and the two artists accuse each other of plagiarism. But they artistically have different *meanings*: while J's painting is devoted to Newton's third law of motion, K's represents the first one. In J's work, the straight line divides two areas, two masses in which the upper mass is pressing down with a force proportional to its acceleration and the lower mass is pressing up with as reaction to the other force. In K's painting, the same line represents the path of an isolated particle which, undisturbed by external forces, moves uniformly in a straight line.

Danto aims at showing how the transfiguration of a commonplace such as a red rectangle, a rectangle crossed by a straight line, or the famous Andy Warhol's Brillo Boxes into a work of art does not depend from material properties or capacities—for instance, the capacity of Warhol and other members of the Factory of reproducing boxes that look exactly like the Brillo Boxes one could find in the supermarket—but rather on the interpretation the artist and the critics give of these materialities: "In art, every new interpretation is a Copernican revolution, in the sense that each interpretation constitutes a new work, even if the object differently interpreted remains, as the skies, invariant under transformation" (125). Danto's perspective on art is hence deeply hermeneutic: in the world of art—but not in the world in general—the motto *"esse est interpretari,"* "to be is to be interpreted," rules for him. But his hermeneutics is not relativistic or nihilistic, in the Nietzschean sense that there are no facts, just interpretations.[3] Indeed, the number of possible interpretations of an artwork is constantly under strict control: legitimate interpretations are those which take place within the limits of art history and theory. This regards both the artist and the critics. The more one knows the history and theory of art, and the more one occupies a central position in the art world, the more she will be legitimate to intervene, by creating or judging, into it. From the perspective of imagination and creativity, such perspective suggests the idea that doing or judging art has less to do with radical novelty than with the possibility and the capacity of orienting oneself in the artworld, and eventually in contributing at reconfiguring some of its elements.

To put it differently, and this is my version of Danto's thesis, the more expertise we gain when it comes to the art world, the more we see the recombinatory nature of a piece of art. This also means that the more we know the art world, the less we have the illusion of pure novelty and creation. Finally, this also implies that the more confidence we have regarding the artworld, the more we are able to discriminate between the repetitive and innovative recombinations, and hence, the more we have the possibility to productively intervene in the art world. Such thesis can be applied, I repeat, to both the aesthetic judgment and the aesthetic production, since art, as Danto points out, is henceforth (or, it has always somehow been) intellectualized, which means that an innovative artist such as Cattelan (whose manual capacities are incidentally far more inferior to his intellectual creativity) does not work in isolation, but always in dialogue and/or in contrast with the past and the present of the art world.

I would like to stress that Danto does not want to propose a universal or eternal theory of art (and this is the reason why his book is modestly subtitled as "*A* [and not "The"] philosophy of art." His theory is valid on the basis of contemporary artworks, but nothing excludes the idea that in a few decades a new artistic paradigm will replace his own. Like with scientific paradigms, new observations can bring us to adapt or invalidate old theories, and to formulate a new theory. This does not exclude, however, that his art theory does not have a certain retroactive validity, in the sense that it can shed light on some aspects of the entire art history. Newton's *Principia Mathematica* were not written in the isolation of a totally bodiless mind, because Newton "was more like a spider in the center of a huge web. [. . .] Newton reaches the stars because he is also the center of a vast empire of information" (Latour 2010, 3). And the same holds true for Leonardo and Michelangelo, who were dealing with their respective worlds, art worlds but also social worlds. Something similar could be said of my theory of imagination-bricolage, which seems to particularly fit with our contemporary 'sample culture,' but also pretends to have a certain degree of long-term validity. This is, for instance, the complex network of authors in which Ricoeur's thinking is entangled (see Figure 4.1).

You probably have already guessed my strategy. Until now, I have just spent a few words on digital technologies and media, and I have focused my attention on human imagination. Instead of exalting digital imagination, I have slightly criticized the human imagination. My inspiring figure has been Bourdieu, and the illuminating pages he wrote in *Outline of a Theory of Practice* (1977, 73–76) against the liberating power Sartre attributes to imagination and against his belief in the possibility of a sort of radical human freedom. According to Bourdieu (1977, 73), "Sartre makes each action a sort of an unprecedented confrontation between the subject and the world." In other words, "he leaves no room for everything that [. . .] might seem to blur the sharp line his rigorous dualism

Figure 4.1 Co-occurence Network Generated by Authors and Author Keywords
Appearing in the Same Paper (Mentioning Ricoeur in the Category
Arts And Humanities Scopus)

Source: Romele and Severo 2016

seeks to maintain between the pure transparency of the subject and the
mineral opacity of the thing" (74–75). I am perfectly conscious of the
risk of appearing a determinist, precisely as Bourdieu has been accused of
being, partly with good reason. I will discuss this point in the conclusion,
and I will consider Bourdieu's notion of *habitus* in the finale.

It is time now to spend some words on digital technologies. In the previous chapter, I resorted to Lev Manovich's notion of software to show
how digital technologies can be understood as imaginative machines, in
the sense of productive imagination. In particular, I have said that to
make an analogy with the productive imagination, one could compare
databases to sensibility and algorithms to the forms of understanding.

The function of an algorithm consists of reorganizing data according to a certain coherence. It is precisely within this material duality that, I believe, one has to understand digital recording and registration. And this is also the reason why, I also believe that digital registration and recording are not passive, but rather imply activity and agency. There is, however, an important difference between *manually* applying a filter on an image on Instagram or *manually* entering a search on the Google search engine for some news (these were two examples done in the previous chapter), and leaving the machine to *automatically* and *autonomously* look up or create the information for us. According to Galit Wellner (2018, 219–221), machines have been writing texts for several decades (and one should not forget the descriptions of authors like Latour and Callon of the machines in laboratories as 'inscriptional' technologies); but according to her, a new type of automatic writing is emerging that comes closer to what we term creativity and until recently classified as uniquely human. This is the case, for instance, of "robo-journalism," which is developed mostly in the areas of financial reporting and coverage of sport events, in which machine learning algorithms scan huge amounts of data and quickly build articles from them. An article recently published on *Wired* speaks about Google's new algorithm that "perfects photos before you even take them."[4] Even without going that far, one can consider Google Photos' assistant, which autonomously proposes stylized photos, animation from videos, and so on, or Google News that suggests news that is supposed to be relevant to you. And one might go on about hundreds of services of this kind offered by Google and other enterprises.

The question arises as to what extent can we say that we are before a creative imagination? On the basis of what I have said before, novelty and creation are mostly observer-dependent. This holds true for both human beings and machines. Certainly, the complexity of the object observed plays an important role. Before it became unplayable, I amused myself with the video game designed by Stefano Gualeni, *Gua-Le-Ni; or, The Horrendous Parade*, based on Hume's notion of complex ideas.[5] *Gua-Le-Ni* took place on the wooden desk of an old British taxonomist. On his desk lay a fantastic book: a bestiary populated by finely drawn creatures. As for the monsters of myths and folklores in general, the impossible creatures in *Gua-Le-Ni* were combinations of parts of real animals. The goal of *Gua-Le-Ni* was that of recognizing the modular components of the fantastic creatures and their relative order before one of them managed to flee from the page. Now, let us imagine that the possible combinations of *Gua-Le-Ni* were not thirty, but thousands or even millions, as numerous and small as the pixels of a smartphone's good camera are. We would not be able to play the game, because we would be unable to recognize the modular components of the fantastic creatures. And we would not recognize that the impossible creatures we see are the result of recombinations of parts of other creatures. They would appear as new creatures. One could certainly say that such recombinations would result in mere chaos. But let us suppose that at the core of *Gua-Le-Ni* is a machine learning algorithm that has been trained with images of what we commonly consider as creatures. The algorithm would then create images that are similar, but not equal, to other creatures we have seen before. We would have brand-new images of fantastic creatures, without any perception of the determinism that lies behind them. Interestingly enough, this example is complex and maybe inappropriate not because of the limits of the machines, but because it questions the limits of our imagination. Can we as human beings really imagine brand-new creatures? Rather, is not our imagination doomed to remain within the limits of anthropomorphism or at least of what has been already seen on Earth? The machine must be fed with our standards, but this not for the machine's own sake, but rather for our needs and fears. The recent case of two of Facebook's AI programs is very well known.[6] The programs were shut down because they appeared to be chatting to each other in a language only they understood.

In a recent article from which I have partially borrowed the title of this section, Manovich (2017) argues that artificial intelligence is playing an important role in our cultural lives, increasingly automating the realm of aesthetics. In particular, he insists on the fact that today AI is not just automating our aesthetic *judgments*, recommending what we should watch, listen to, or wear. It also plays an important role in some areas of aesthetic *production*. AI is used to design fashion, logos, music, TV commercials, and other products of the culture industry (3). For sure, in most cases, it is still fair to speak of a data-*driven* culture, in the sense that human beings still make the final decision. For example, this is the case with *Game of Thrones*, in which the computer suggests plot ideas but the actual writing is done by humans. This is also the case of what has been called the first "AI-made movie trailer," that of the sci-fi thriller *Morgan*, made with the help of IBM Watson. The computer selected many shots suitable to be included in the trailer, and then a human editor made the work. But it is only a matter of time before entirely AI-designed cultural products will be available on the market of the culture industry; and actually, they are already widely available in the art world. Similar remarks can be found in Manovich (2018, 2–3): "I expect that production and customization of many forms of at least 'commercial culture' [. . .] will also be gradually automated. So, in the future already developed digital distribution platforms and media analytics will be joined by the third part: *algorithmic media generation* (of course, experimental artists, designers, composers, and filmmakers have been using algorithms to generate work since the 1960s, but in the future, this is likely to become the new norm across culture industry)." Manovich (2017, 6) also stresses how the use of AI and other computer methods is helping us find regular patterns in culture, such as influence between artists, changes in popular music, and gradual changes in average shot duration across thousands of movies. According to him (Manovich 2018, 7), "the point of any application of quantitative or computational methods to analysis of culture is not whether it ends up being successful or not (unless you are in media analytics business).[7] It can force us to look at the subject matter in new ways, to become explicit about our assumptions, and to precisely define our concepts and the dimensions we want to study." This means that digital machines are also teaching us to be modest when it comes to us pretending to be engineers.

* * *

The main advantage of the approach I have developed in this chapter is that I do not have to position myself among the techno-enthusiasts or the transhumanists in order to defend my principle of symmetry between humans and machines. The biggest disadvantage, however, is to appear, as I have already said, a determinist. I have no difficulty in recognizing myself as a defender of social determinism. In this case, I would even define myself as a staunch defender of determinism, as staunch as the Foucault of the first volume of *The History of Sexuality*, in which he describes the process of interiorization of the obligation to confess (*avouer*):

> The obligation to confess is now [. . .] so deeply ingrained (*incorporée*) in us, that we no longer perceive it as the effect of a power that constrains us; on the contrary, it seems to us that truth, lodged in our most secret nature, 'demands' only to surface [. . .] Truth does not belong to the order of power, but shares an original affinity with freedom.
>
> (Foucault 1978, 60)

In social life, freedom is no more than a vanishing point. The same holds true, for instance, for the creation of an artwork if seen from a specific art world perspective à la Danto. Neither do I have difficulty in recognizing myself as a determinist whenever it is a matter of mediation and exteriorization, such as in technology, language, and embodiment. And yet, I have some difficulties in affirming that determinism is the ultimate human condition. Maybe this is just a case of residual human conceit. And maybe from a higher perspective, that of a god or a superior machine, we would appear as pure matter, with all our behaviors already inscribed in our (genetic?) code. All I know is that as our knowledge stands at present, we can still *believe* in a tension, although minimal, between auto- and hetero-determination in human beings. I will further discuss this point in the finale.

Notes

1. It must be noted that Estival (2002, 153–159) criticizes the perspective of Jean Bertin, the father of the information visualization in France, for his "endo-genic" perspective of the graphic schemas, that is to say the general lack of interdisciplinarity and interest for the intrinsic relation between internal and external schemas.
2. M. Kosoff, "Trump Wants a 'Transparent' Border Wall to Prevent Injuries from Falling 'Sacks of Drugs.'" *Vanity Fair*, July 13, 2017, https:// www.vanityfair.com/news/2017/07/trump-transparent-border-wall-falling-drugs-mexico. Accessed June 10, 2019.
3. In Gadamer as well art is intrinsically hermeneutic. The difference from Danto is, however, radical. For both of them, in the artworld "*esse est interpretari.*" And yet, while for Danto the work of art is the object of a human (depend-ing on the artworld) interpretations, for Gadamer the contrary seems true. According to the German philosopher (Gadamer 2004, 102–130), in the authentic art experience, human beings are taken, or interpreted, by the sole truth-event that lies within, or behind, or at the core of a true work of art.
4. E. Stinson, "Google's New Algorithm Perfects Photos Before You Even Take Them." *Wired*, July 8, 2017, www.wired.com/story/googles-new-algorithm-perfects-photos-before-you-even-take-them/. Accessed June 10, 2019.
5. https://en.wikipedia.org/wiki/Gua-Le-Ni;_or,_The_Horrendous_Parade. Accessed June 10, 2019.
6. A. Griffin, "Facebook's Artificial Intelligence Robots Shut Down Because They Start Talking to Each Other in Their Own Language." *Independent*, July 31, 2017. www.independent.co.uk/life-style/gadgets-and-tech/news/face book-artificial-intelligence-ai-chatbot-new- language-research-openai-google-a7869706.html. Accessed June 10, 2019.
7. It is noteworthy that in this article Manovich considers Bourdieu's quantita-tive analysis of cultural data in the *Distinction* as an antecedent of his digital cultural analytics. See the finale for some considerations on the resemblances between Bourdieu's sociology and methodology, and the new algorithmic machines.

References

Bachimont, B. 2010. *Le sens de la technique: le numérique et le calcul*. Paris: Les Belles Lettres.

Beaubois, V. 2015. "Un schématisme pratique de la raison." *Appareil* 18. https://appareil.revues.org/2247?lang=en. Accessed June 10, 2019.

Black, M. 1962. *Models and Metaphors*. Ithaca: Cornell University Press.

Bourdieu, P. 1977. *Outline of a Theory of Practice*. Cambridge: Cambridge University Press.

Danto, A. 1981. *The Transfiguration of the Commonplace: A Philosophy of Art*. Cambridge, MA: Harvard University Press.

De Certeau, M. 1984. *The Practice of Everyday Life*. Berkeley: University of California Press.

De Mul, J. 2018. "Encyclopedias, Hive Minds and Global Brains: A Cognitive Evo-lutionary Account of Wikipedia." In A. Romele, and E. Terrone (eds.). *Towards a Philosophy of Digital Media*. London: Palgrave Macmillan, 103–118.

Derrida, J. 1978. *Writing and Difference*. Chicago: The University of Chicago Press.

Estivals, R. 2003. *Théorie générale de la schématisation, tome 3. Théorie de la communication*. Paris: L'Harmattan.

———. 2003. *Théorie générale de la schématisation, tome 1: Théorie de la communication*. Paris: L'Harmattan.

———. 2002. *Théorie générale de la schématisation, tome 2: Sémiotique du schéma*. Paris: L'Harmattan.

———. 2001. *Théorie générale de la schématisation, tome 1: Épistémologie des sciences cognitives*. Paris: L'Harmattan.

Ferraris, M. 2012. *Documentality: Why It Is Necessary to Leave Traces*. New York: Fordham University Press.

Flusser, V. 1999. *The Shape of Things: A Philosophy of Design*. London: Reaktion Books.

Foucault, M. 1978. *The History of Sexuality: Volume 1: An Introduction*. New York: Pantheon Books.

Gadamer, H.G. 2004. *Truth and Method*. London and New York: Continuum.

Hills, A., and Bird, A. 2018. "Against Creativity." *Philosophy and Phenomenological Research*. Online first. https://onlinelibrary.wiley.com/doi/full/10.1111/phpr.12511.

Kant, I. 1987. *Critique of Judgment*. Indianapolis and Cambridge: Hackett Publishing Company.

Latour, B. 2010. "Networks, Societies, Spheres: Reflection on an Actor-Network Theorist." Keynote Speech for the International Seminar on Network Theory: Network Multidimensionality in the Digital Age: Annenberg School for Communication and Journalism, Los Angeles, February 19, 2010. www.bruno-latour.fr/sites/default/files/121-CASTELLS-GB.pdf. Accessed June 10, 2019.

Lévi-Strauss, C. 1966. *The Savage Mind*. London: Wiedenfeld and Nicolson.

Maddalena, G. 2015. *Philosophy of Gesture: Completing Pragmatists' Incomplete Revolution*. Montreal and Kingston: McGill-Queen's University Press.

Manovich, L. 2018. "Can We Think Without Categories?" http://manovich.net/index.php/projects/can-we-think-without-categories. Accessed June 10, 2019.

———. 2017. "Automatic Aesthetics: Artificial Intelligence and Image Culture." http://manovich.net/content/04-projects/101-automating-aesthetics-artificial-intelligence-and-image-culture/automating_aesthetics.pdf. Accessed June 10, 2019.

Mélice, A. 2009. "Un concept lévi-straussien déconstruit: le 'bricolage'." *Les temps modernes* 5(265): 83–98.

Ricoeur, P. 2004. *The Conflict of Interpretations: Essays in Hermeneutics*. London and New York: Continuum.

Romele, A., and Severo, M. 2016. "From Philosopher to Network: Using Digital Traces for Understanding Paul Ricoeur's Legacy." *Azimuth* 7: 113–128.

Simondon, G. 2014. *Imagination et invention (1965–1966)*. Paris: P.U.F.

Wellner, G. 2018. "From Cellphones to Machine Learning: A Shift in the Role of the User in Algorithmic Writing." In A. Romele, and E. Terrone (eds.). *Towards a Philosophy of Digital Media*. London: Palgrave Macmillan, 205–224.

Conclusion

In the introduction to the second part of this book, I have accounted for the anthropocentrism that has characterized the history of classic hermeneutics. But as I have said, during the second half of the twentieth century, some researchers have also made the effort to go beyond this perspective, while remaining within a hermeneutic framework. In particular, some authors have been interested in the question of nature's hermeneutics. In this context, two trends can be distinguished. The first is that of those who have argued that nature as well could be considered as an object of interpretation. This is the case of environmental hermeneutics, which is interested in the way in which the representations we have of nature determine our relationship with it. In fact, as Forrest Clingermann clearly observes, the peculiarity of an environmental hermeneutics is that it cannot be reduced either to social constructivism (i.e. to a mere projection of our assumptions and preconceptions on it) or to scientism (i.e. a mere 'objective' and 'distant' observation of what is there before us). For an environmental hermeneutics, "[n]ature has independent existence that can be explained, but it is also influenced and understood in light of the observer's perspective and experience" (Clingermann 2013, 249). In short, if on the one hand human beings give nature certain meanings depending on different factors such as the social and cultural contest, on the other hand, nature itself has, so to speak, its affordances. As Martin Drenthen writes in this regard (2015, 167) "hermeneutics believes that moral meaning exists within human understanding, but the process of interpretation is not a process of *constructing* but rather of *responding* to an experience of meaning." In a certain sense, it is precisely the affordances of nature that can help us in distinguishing between good and bad interpretations.[1]

If most of the hermeneutics of nature of this kind carry on specific places or environments, recently the francophone philosopher Alexander Federau (2017, 339–384) has proposed a "planetary hermeneutics." At the time of the Anthropocene, a double mediation at the planetary level is at work for him. On the one hand, the whole Earth becomes a medium that records signs. Geology is already a science of the trace, but the traces of the Anthropocene are different since they have human origins. The Anthropocene is the moment in which geological history and human history meet. On the other hand, just as Clingermann says, we have symbolic forms, imaginaries, and mediating images, which in this case are no longer partial or local, but represent the Earth as a whole. The author does not refer to *The Blue Marble*, the image of planet Earth made in 1972 by the crew of Apollo 17 that recalls the fragility and beauty of our planet; he rather refers to *The Black Marble*, a picture of Earth composed of data obtained by the Visible/Infrared Imager Radiometer Suite instrument onboard the Suomi NPP satellite in 2012. In it, one can see the artificial lights of some of the great metropolitan areas of the world, which remind us how humans have become a force on a global and geological scale. He also speaks of comparative images, such as those proposed by the Nasa *Images of Change* project.[2]

The second tendency in the field of the hermeneutics of nature is that of those who have argued that nature can be seen as a proper subject of interpretation. The cost of this radicalization is a certain reduction, because in this case the hermeneutics of nature can concern the organic life alone—with a certain predilection for the animal world.

In this regard, two strategies have been developed to divert the problem. The first is an 'internalistic' strategy, conceived for example in the context of biosemiotics, and which consists in distinguishing between different levels of signification within a single process of semiosis. In the case of single cells, one can then speak for example of "manufacturing semiosis," which is a form of coding and decoding. As Marcello Barbieri (2009, 235–236) writes, rather than bringing biology to humanities, it is about keeping it within science, "because it [meaning] is a natural entity and we must introduce it into science just as we have introduced the concepts of energy and information." According to the Dutch philosopher Jos De Mul (2016), three postulates characterize biosemiotics: (1) all life forms are characterized by semiosis, i.e. processes, activities or conduct involving the production and elaboration of codes, signals, or signs; (2) semiotic elements such as codes, signals, and signs and their coding, decoding, reading or interpreting are natural phenomena; (3) life is characterized by an emergent evolutionary history, in which the semiosis becomes increasingly more differentiated and complex. He also describes this process of differentiation and complexification in nature as a transition from the indexical level of a cell (e.g. when a catalyst causes a chemical reaction in an organ), to the iconic one of the animal (e.g. the dog recognizing certain movements as playful behavior and seeing in them an invitation to play), to, finally, the symbolic level of human beings and of some animals endowed with language, which is also the properly hermeneutic level.

The other strategy consists of following an "externalist" path, in the sense of being interested in the phenomena of appearing/appearance and of the "address." Jean-Claude Gens (2008) refers, for example, to the comprehensive biology of Portmann and to his notion of "signal" (*Merkmal*). Portmann's works, especially from the mid-fifties, bring about the "sense of manifestation" or "appearance" (*der Sinn der Erscheinung*) of the living form, that is, its expressive (*Ausdruckswert*) or (re)presentative (*Darstellungswert*) value. In short, it is not about studying animal behavior or its evolution, but the multiplicity of its forms (and its drawings, colors, et cetera.), against the dominant teleologism in Darwinian and neo-Darwinian thought. From this comes the concept of *Merkmal*, which is "a distinctive sign or signal in a non-communicational [. . .] sense that escapes the study of animal interiority, or the internal processes of the organism" (190). This animal and plant aesthetics cannot be entirely reduced to the needs of reproduction, selection, and evolution. For Portmann as for Gens, it is a matter of studying and listening to this "surface language," without looking for intentions and purposes, because, as Portmann quoted by Gens (193) says, "a thing without purpose is not necessarily a thing without meaning" (*zwecklos is nicht sinnlos*). The French philosopher also refers to the eulogy of the appearance of Hannah Arendt, who recalls some of Portmann's theses in her latest book, *The Life of the Mind*.

Summarizing what has been said up to now, we can distinguish three levels in the hermeneutic approach to nature: (1) a 'zero' level, in which nature cannot be interpreted and, in any case, nature does not interpret by itself; (2) a level 'one,' in which nature is the object, not entirely passive, of our projections and expectations of meaning; (3) a level 'two,' which dares to think about the interpretative capacities of nature itself, according to an internalist or externalist perspective. My thesis is that the same three levels can be found in the emerging field of digital hermeneutics:

1. First of all, there is a level 'zero,' in which hermeneutics (especially the Heideggerian one) has been used to mark a clear distinction between humans and non-humans (machines). According to Hubert Dreyfus, there is an essential difference between human beings and computers: "[t]he human world, then, is prestructured in terms of human purposes and concerns in such a way that what counts as an object, or is significant about an object already, is a function of, or embodies, that concern. This cannot be matched by a computer, for a computer can only deal with already determinate objects" (Dreyfus 1972, 173). Human beings have goals, which are realized on the basis of a system of values and emotional states that are usually not explicit. Machines, instead, have ends, which are rather realized (Dreyfus is referring here to symbolic AI) thanks to a predefined list of specific criteria. Even in more recent publications, he has insisted on such an intrinsic difference between humans and machines, by denouncing, for instance, the insufficiency of the attempts to build

a "Heideggerian AI." The fact is that, for him, we would need "a model of our particular way of being embedded and embodied. [. . .] That is, we would have to include in our program a model of a body very much like ours" (Dreyfus 2007, 1160). It is interesting to notice how Dreyfus's Heideggerian radicalism is in this context more radical than Heidegger himself. He refuses, for instance, the notion of as-structure, which plays an important role in *Being and Time*. For him, "when absorbed in coping [with the world], I can be described objectively as using a certain door as a door, but I'm not experiencing the door as a door" (1141). In other words, the as-structure is already a derivative way of understanding our dynamic and immediate coping with the world. To put it a bit differently, one might argue that from Dreyfus's perspective, Heideggerian "readiness-to-hand" (*Zuhandenheit*) is still a form of "presence-at-hand" (*Vorhandeinheit*).

What surprises me most about this perspective is not so much what Dreyfus thinks of machines. His criticism of GOFAI (good old-fashioned artificial intelligence) is indeed right, and his opinions are today widely shared among scholars, both philosophers and engineers. Rather, it is surprising what he says or, rather, does not say about human beings. In fact, while Dreyfus's theses on machines are based on those that in his time were the most advanced achievements in the field, his considerations of human beings are still based on the thesis that Heidegger, an inhabitant of a small town on the edge of the Black Forest, had thought in the twenties. But one wonders if those theses are still valid today for us living in or at the margins of the great metropolises of the world. One wonders, above all, if the innocence of the Heideggerian human being, who lives in a relationship of immediacy with the world, has not now been definitively shattered. And one wonders how the relationship with machines (and the prolonged contact with sociology, critical theory, irony, cynicism, old and new media, and all other forms of self-distanciation and self-analysis) has determined precisely this rupture. Are not our actions and interactions with the world, our appearances and our gestures, increasingly thought through? The fact is, as trivial as it is true, that by creating digital machines we have also begun to interpret and shape ourselves in their image and likeness. Dreyfus almost says it at the end of *What Computer Can't Do* (1972, 192): "Man's nature is indeed so malleable that it may be on the point of changing again. If the computer paradigm becomes so strong that people begin to think of themselves as digital devices [. . .] then [. . .] human beings may become progressively like machines." But he does not really seem to draw any consequence from this strong statement.

2. Then there is a level 'one,' in which the interpretation is a result of the articulation between human and non-human intentionalities. For Capurro (2000, 80), according to existential hermeneutics, a human

inquirer is not an isolated system trying to reach others from her encapsulated mind/brain. Rather, she is always-already sharing the world with others. Similarly, "the information-seeking process is basically an interpretation process having to do with the (life-)content and the background of the inquirer and with that of the people who store different kinds of linguistic expressions having a meaning within fixed contexts of understanding (as, for instance, thesauri, keywords and classification schemes)."

More recently, Capurro (2010) affirmed that what is new with regards to digital hermeneutics are two sides of a single weakening process of modern technology. On the one side, there is the weakening of the interpreter that finds herself within a network of human and non-human actants that she cannot really control. On the other side, information technology is a weak technology, since it deals with human conversations. The digital and hermeneutics are then entangled: the former questions the autonomy of the interpreter, and the latter seems able to analyze and eventually reconfigure the structure of the technological system. From an anthropological perspective, then, digital hermeneutics questions the interpretational autonomy of human beings, our general loss of control on the way we interpret, and hence, see the world.

Such general loss of control due to digital technologies has been seen both positively and negatively. Recently, Sützl (2016) has highlighted this tension in Gianni Vattimo's hermeneutics of media. Earlier works of the Italian philosopher suggest that we should embrace the aestheticization of experience caused by interacting with digital media and technologies. Indeed, for him, mass media renders society complex and chaotic, and it is precisely in this "relative chaos" that our possibilities of emancipation lie. The same could hold true today for digital media. And yet, over the course of the years, Vattimo has become more and more sensible to the anesthetizing effects of media. From his more recent works, one can deduce the idea according to which "the dominant digital media are not capable of furthering processes of emancipation that are anything more than a liberal model of emancipation and as such can be utilized for furthering the neoliberal neutralization of politics" (755).

3. Finally, there is a level 'two' that still has mainly to come, and that would consist of wondering if it is legitimate, and eventually to which extent, to attribute to digital technologies, or at least to an emerging part of them, an autonomous interpretational agency. Incidentally, it is noteworthy that at this level we assist in a continuous overlapping between an internalist and externalist strategy. On the one hand, the analysis of the machine's emerging properties remains at the surface of the effects and the products—consider, for instance, the Turing Test for artworks, in which people are asked to say which work of art is made by a computer and which one is made by a human

(Boden 2010). On the other hand, the same analysis often tries to go beyond appearances. The debate between symbolism and connectionism is a good example. The recent success of artificial neural networks has given new force to the internalist tendencies.

In "Making a Mind Versus Modelling the Brain" (1988) Hubert Dreyfus and Stuart Dreyfus have proposed an interesting analogy between hermeneutics and connectionism. According to them, the symbolic AI and the rationalist tradition it refers to have failed. On the contrary, "Frank Rosenblatt's [the inventor of the perceptrons] intuition that it would be hopelessly hard to formalize the world and thus give a formal specification of intelligent behavior had been vindicated" (34). According to them, "neural networks may show that Heidegger, later Wittgenstein and Rosenblatt were right in thinking that we behave intelligently in the world without having a theory of that world" (35).

Does it mean that neural networks are actually similar to human minds? For the authors, the answer is negative. First of all, because human beings have bodies, needs, emotions, et cetera, they are then much more holistic in nature than neural nets. Second, and this is what most interests me in this context, artificial neural networks' abstractions are not by force similar to human abstractions. If this does happen, it is just to make the machines' choices understandable to us. In the words of the Dreyfus brothers, "[t]he designer of the net has a specific definition in mind of the type required for a reasonable generalization, and counts it a success if the net generalizes to other instances of this type. But when the net produces an unexpected association can one say it has failed to generalize?" (38). Similarly, one might ask: when an unsupervised machine learning algorithm produces an unexpected abstraction, association, or correlation, could not we say that we are still standing before a form of interpretation, however not human anymore? In the field of digital art and aesthetics, Al-Rifaie and Bishop (2015) have proposed to distinguish between weak and strong computational creativity. While the former does not go beyond imitating the human creativity, the latter leaves the machine free to express its own cognitive states.[3] Can such a distinction be applied to domains other than art and aesthetics? Floridi and Sanders (2004) have spoken of "aresponsible morality" for machine learning algorithms; Katherine Hayles (2014) has introduced the notion of "nonconscious cognition." In a similar way, could one speak for certain digital machines of interpretation without proper understanding?

Notes

1. John van Buren (2013, 28–34) introduces more precisely four criteria for assessing interpretations of the environment and of the forest in particular: (1) the biophysical criterion (interpretations must fit the biophysical world to

which they refer); (2) the historical criterion (interpretations must fit historical and social traditions—do they involve "creative growth" or rather "alienating disruption"?); (3) the technical or pragmatic criterion (Does the interpretation work? Is a certain perspective efficient in relation to the end in view that is to be produced?); (4) the communicative ethical-political criterion (here one asks if the adoption of the public sphere of certain views, values, and politics has been or can be arrived at in conformity with the fundamental procedural norms or ideals for making decisions democratically in society). Clearly, "[n] one of the criteria on its own gives us the whole truth. The ideal here would be to satisfy all four criteria or find a creative compromise between them" (34). Only the last criterion has a sort of meta-function with respect to others because it applies both to the political sphere and to the democratic balance between the other criteria.

2. https://climate.nasa.gov/images-of-change?id=652#652-flooding-and-fire-in-botswanas-okavango-delta. Accessed June 10, 2019.
3. For a more nuanced perspective, see Coeckelbergh (2017).

References

Al-Rifaie, M.M., and Bishop, M. 2015. Weak and Strong Computational Creativity. In M. Schorlemmer, A. Smaill, and T. Besold (eds.). *Computational Creativity: Towards Creative Machines.* Dordrecht: Springer, 37–49.

Barbieri, M. 2009. "A Short History of Biosemiotics." *Biosemiotics* 2: 221–245.

Boden, M.A. 2010. "The Turing Test and Artistic Creativity." *Kybernetes* 39(3): 409–413.

Capurro, R. 2010. "Digital Hermeneutics: An Outline." *AI & Society* 25(1): 35–42.

———. 2000. "Hermeneutics and the Phenomenon of Information." In C. Mitcham (ed.). *Metaphysics, Epistemology, and Technology: Research in Philosophy and Technology, Volume. 19.* Amsterdam: Elsevier, 79–85.

Clingermann, F. 2013. "Memory, Imagination, and the Hermeneutics of Place." In F. Clingermann, B. Treanor, M. Drenthen, and D. Utsler (eds.). *Interpreting Nature: The Emerging Field of Environmental Hermeneutics.* New York: Fordham University Press, 245–263.

Coeckelbergh, M. 2017. "Can Machines Create Art?" *Philosophy & Technology* 30(3): 285–303.

De Mul, J. 2016. "The Syntax, Pragmatics and Semantics of Life: Reading Dilthey in the Light of Contemporary Biosemiotics." In C. Damböck, and H. U. Lessing (eds.). *Dilthey als Wissenschaftsphilosoph.* Freiburg: Verlag Karl Alber, 156–175.

Drenthen, M. 2015. "Environmental Hermeneutics and the Meaning of Nature." In S. M. Gardiner, and A. Thompson (eds.). *The Oxford Handbook of Environmental Ethics.* Oxford: Oxford University Press, 162–173.

Dreyfus, H. 2007. "Why Heideggerian AI Failed and How Fixing It Would Require Making It More Heideggerian." *Artificial Intelligence* 171(18): 1137–1160.

———. 1972. *What Computers Can't Do.* Cambridge: The MIT Press.

Dreyfus, H., and Dreyfus, S. 1988. "Making a Mind Versus Modeling the Brain: Artificial Intelligence Back at a Branchpoint." *Daedalus* 117(1): 15–43.

Federau, A. 2017. *Pour une philosophie de l'anthropocène.* Paris: P.U.F.

Floridi, L., and Sanders, J. 2004. "On the Morality of Artificial Agents." *Mind & Machines* 14(3): 349–379.

Gens, J-C. 2008. *Éléments pour une herméneutique de la nature*. Paris: Éditions du Cerf.

Hayles, K. 2014. "Cognition Everywhere: The Rise of the Cognitive Nonconscious and the Costs of Consciousness." *New Literary History* 45(2): 199–220.

Sützl, W. 2016. "Gianni Vattimo: Hermeneutic Communism and Digital Media Theory." *Philosophy Today* 60(3): 743–759.

van Buren, J. 2013. "Environmental Hermeneutics Deep in the Forrest." In F. Clingermann, B. Treanor, M. Drenthen, and D. Utsler (eds.). *Interpreting Nature: The Emerging Field of Environmental Hermeneutics*. New York: Fordham University Press, 17–35.

Finale
The Indifferent Ones

At the end of the second part of this book, I plead for a "posthuman hermeneutics," which should overcome the anthropocentrism that characterized most of the history of this discipline. I have long wondered if those remarks should also be the conclusion of this book: a sort of 'happy ending' then, a radical principle of symmetry through which digital machines find their dignity in a harmonious community between humans and non-humans. Finally, I have decided to take a step forward that, at a first glance, might be perceived as half a step back.

To actualize digital hermeneutics according to the level 'two' I have presented, three series of considerations are indeed necessary:

1. First of all, from an internalist perspective, digital hermeneutics should be able to distinguish between different degrees and kinds of interpretation. Hans Lenk's distinction between six different levels of interpretation I have presented in chapter 3 might represent an interesting framework. Digital hermeneutics should establish which of these levels are or could be soon implemented into digital machines, and which remain instead a human prerogative. For example, whereas it is clear that digital machines have exceeded human capacities in terms of classification, they still face difficulties in competing with human pattern-recognition abilities. Moreover, one could wonder if justification—interpretation and meta-interpretation—could ever be fully implemented in digital machines. Digital hermeneutics should also investigate those emerging forms of digital interpretation that have little in common with human ways of coping with the world;

2. Second, from an externalist perspective, digital hermeneutics should deal with the degree of mutual understandability between humans and digital machines. For instance, while some outputs of these machines—along with the process that led to them—are humanly interpretable and understandable, many others are not. The interpretational gap between humans and digital machines can have no important consequences in fields such as art and cultural production

in general. Yet what happens when such a gap manifests itself in other contexts like justice, politics, and economy?

3. Third, digital hermeneutics cannot be identified with a joyful plea for a generalized principle of symmetry. Critical posthumanism is engaged, for instance, in *prescribing* some actions and some choices, in assuming the responsibility of preferring some modes of existence to others. It is for this reason that Braidotti (2016, 24) affirmed that "[p]osthuman critical thought is not post-political. The posthuman condition does not mark the end of political agency, but a re-casting of it in the direction of transversal alliances and relational ontology." In other words, the task of a critical posthumanism is to find differences albeit while remaining within the limits of a principle of symmetry. This same attitude must be encouraged in the field of digital hermeneutics, and more widely in all hermeneutics that wants to go beyond the limits of anthropocentrism.

In the first part of this book, I defended the idea that 'the virtual never ended.' In particular, I have taken a position against those who more or less consciously believe, from an ontological or epistemological perspective, in a strong homology between online and offline. In the second part of the book, I argued that some digital machines have emerging properties that increasingly resemble human imaginative and interpretative capacities. I also claimed that human beings could find inspiration in these machines in order to lower some of their pretentions in terms of creativity and authenticity. Is this contradictory? Have not I proposed two opposite digital hermeneutics, the first insisting on limits and differences, rather the second stressing the resemblances between humans and digital machines?

A possible answer to this potential critique is that digital hermeneutics can only be conflictual in nature. There will always be defenders of the human 'touch,' just as much as there will always be an attraction for the 'cold' role of the digital in the interpretative processes. Incidentally, one should not forget that hermeneutics has been defined as being *both* an art and a technique or technology. This implies two sorts of elements or competences. Traditionally, the peace between these two factions has never lasted for long. This is also the case of digital hermeneutics, despite all efforts, from me as well as from other authors, to keep humanist and technicist tendencies together. Thus should I conclude this book with a plea for the conflict of rival interpretations in this emerging field? I don't think this is plausible. In a certain sense, pleading for conflict or war as the "father and king of all" has often been another easy and joyful compromise in philosophy. Rather, the moment has come for discerning, discriminating, and deciding between competing claims in meaning. In other terms, instead of being content with the contradiction, I propose to go through what Kearney (2011) has called "diacritical hermeneutics."[1]

Beyond provocation, it is important to insist for instance on the fact that *e*magination has, as its starting point, data abstracted from their origin and history. It is also noteworthy that while human imagination-bricolage mostly operates in time (memories, stories, traditions, et cetera), in digital machines it is always a matter of space. An algorithm is a structure of spatialization, and even time is algorithmically spatialized.

The main question I would like to ask in this finale is about the kind of interference between human and non-human (in this case, digital) claims for meaning when the object of interpretation and eventual understanding is human subjectivity. My idea is that this is one of the places in which the need for decision making and prescription emerges with the strongest of urgencies. It is not just for the sake of rhetoric that I would also contend that the interference between humans and digital machines, when it comes to interpreting and understanding subjectivities, has today taken the form of indifference. For me, indifference is the most important affection in the present digital age.

The Indifferent Ones (also translated as *The Time of Indifference*) is the title of a 1929 novel by the Italian novelist Alberto Moravia. The story is about an impoverished bourgeois family: Carla, Michele, and their mother, Mariagrazia, and their passive adaptation to the bourgeois clichés, models, and ways of existence in Fascist Italy. Carla and Michele are two young adults who are unable to feel any emotions and thus find themselves at the mercy of boredom and indifference before the social and economic decline of their family. In their actions and decisions there is no rebellion, but just resignation and acceptance. Even when Michele, in the last part of the book, discovers that Leo, their mother's lover who in reality just wants to take possession of the family mansion, has seduced his sister, indifference prevails. Michele decides to challenge Leo to a duel, not because he really feels insulted, but for mere convention. The indifference Moravia represents is a form of apathy, which is generally defined as a lack of feeling, emotion, interest, and concern. But such apathy is very different from that which is thought by philosophers, Stoics for instance. For them, *apatheia* was a means to obtain better knowledge and control of oneself (especially of one's own passions), while Moravia's indifference is precisely toward oneself, and one's own beloved ones. Even when, as for Church Fathers and mystics, *apatheia* became a principle of self-emptying in order to make room for God, salvation remained the ultimate goal. Different is the case of Moravia's characters, who rather seem to embrace a sort of self-perdition.

My intention in this finale is to investigate the indifference in the age of Big Data and the new algorithmic machines, which are the two 'ingredients' of the contemporary digital structure and culture. Such indifference takes, according to me, two directions. First, there is the indifference of the digital toward the human subjects because it works at a supra- or infra-individual level. This does not mean, however, that the digital has no consequences on these subjects. This is precisely the other direction the indifference that interests me takes. The indifference which Big Data and new algorithmic machines have toward subjects is transposed into an indifference subjects have toward themselves and their individualities. To put it differently, the digital is mainly characterized by personalizations without individuation.

In particular, I am going to resort to Pierre Bourdieu's notion of *habitus*. Such notion precisely suggests that (1) there are apparatuses that are radically indifferent to subjects, and that support the reproduction of social and class distinctions; (2) subjects apply these forms of reproduction to themselves. After a certain period of reiteration and internalization, the *habitus* becomes so cognitively and bodily embedded into them that it becomes the prism of all their possible decisions and desires.[2] My hypothesis is that the digital, as it is today, must be considered as a sort of *habitus* generator.

The first argument I would like to develop is that the digital in its current form is less Latourian than Bourdieusian. What the two authors have said about the social reality can be 'translated' into the digital—the biggest difference between them being that while Latour, as I have showed in section 2.2, has operated such translation by himself, Bourdieu, who died in 2002, has never reflected on it.

On September 15, 1998, an article by Latour appeared in *Libération*, the center-left newspaper founded by Jean-Paul Sartre in 1973. The article was significantly entitled "In order to be leftist, it is not enough to talk about the dominated. Nor to give moral lessons to the powers in the name of a sociology. Does the left really need Bourdieu?"[3] I would like to quote a rather long passage of it:

> The sociology of Bourdieu, after a moment of often remarkable description, replaces the multiplicity of terms and situations with a small number of notions, always repeated, and which describe the invisible forces by which the actors are not aware of being manipulated. However, there is an essential difference between the terms invented by the people themselves to define these invisible forces and the "invisibles" revealed by the sociologist: the first are elaborated by the actors and they can "deal" with them; the latter, known only to the sociologist, escape people. Once the dominant discourse of domination has passed, are not ordinary people reduced, even more so, to impotence?

Bourdieu's sociology is, according to Latour, scientistic, because (1) like a biologist or a neurologist, he reveals dynamics of which the social actors are not aware and (2) he looks for a finite ensemble of rules and 'ingredients' in order to explain his object of study, the social reality. But Bourdieu does not proceed in a really scientific way, since (1) he does not care about finding exceptions to his rules, as it would be the case according to the principle of falsifiability, and (2) he does not even care about individuating ways to deal with and eventually modify these hidden forces.

In section 2.2, I said how for Latour the idea that in sociology one has to distinguish between at least two levels, the level of the individuals and the level of the institutions, class tendencies, et cetera, is merely the result of a weakness of the discipline in terms of methods and instruments. I have quoted a passage of *Reassembling the Social* in which Latour states: "it's if as we had to emulate in social theory the marvelous book *Flatland*, which tries to make us 3-D animals live inside a 2-D world only made up of lines. It might seem odd at first but we have to become the Flat-Earthers of social theory" (Latour 2005, 171–172). Latour has limited his sociology to a series of ethnographic micro-observations of the social, plus, of course, a minimal theory of social reality as a flatland. Bourdieu, on the contrary, tried to develop a theory of practice, but he also insisted on the fact that such a theory must be for the practice, and not for its own glory. For this reason, he took a position for instance against the "scholastic epistemocentrism," a certain way of understanding and doing philosophy which considers a metadiscourse at the origin of discourse, or a metapractice at the origin of practices: "This is the 'philologism' which [. . .] tends to treat all languages as dead languages; [. . .] it is the intellectualism of the structuralist semiologists who treat language as an object of interpretation or contemplation rather than an instrument of action and power" (Bourdieu 2000, 52–53).[4]

My first concern with Latour is that I do not see why his micro-perspective should be more leftist than the Bourdieusian attempt to find regularities behind the actions and interactions of social actors. There is a form of elitism in Latour's refusal of standard, quantifiable, and ultimately shareable methodologies. Mapping controversies, for instance, cannot but be complex and never exhaustive. Such an approach will be always in search for the 'intuition' of the 'genius'—for a similar critique, see Pinto (2009). It is true that Bourdieu is skeptical about the possibility and the utility of giving back the word to the dominated. Indeed, for him, the dominated tend to repeat discourses the dominant elites make of them. But he also insists on the fact that this does not mean that sociologists have to take the word *in the place of* the dominated. Rather, he believes in a 'collaboration within separation' between those intellectuals who are specialists in explication and the dominated. The former give to the latter the necessary instruments in order to undertake a symbolic work of political representation and mobilization, in a sort of "transference of cultural capital" (Emmenegger 2016, 131).

My second, and most important in this context, concern about Latour's perspective concerns the kind of social ontology he pleads for by renouncing Bourdieu's "invisible forces," and ultimately all forms of distanciated representation. I would say that Latour's social ontology is an ontology of immediacy and presence. Such a critique is undoubtedly counterintuitive, especially if one considers Latour's predilection for (material) relations over substances. But I also believe that he remains within the limits of a metaphysics of presence insofar as he privileges the visibility of materialized translations and transcriptions over the invisibility of imaginaries and symbolic forms (although, of course, not all symbolic forms are invisible). Not to mention the fact that by trying to develop a micro-approach that never takes distances, and never makes proper abstractions, the risk of getting lost among the actants and their relations is great. It is my conviction that this is a risk related to all the philosophies of technology of the "empirical turn."

I have already expressed my doubts about Latour's understanding of social reality and social science, and have already discussed the limits of his comprehension of the digital as the most adequate mean for approaching social reality. For him, the digital has two properties: (1) it offers a series of instruments and methodologies that finally allow studying social reality without abstractions and simplifications; (2) it, and more specifically its version 2.0, is the adequate paradigm for seeing social reality as it is, that is as an actor-network. The abundance of digital traces and the development of appropriate methods to collect, analyze, and visualize them allows us to overcome that eternal gap between hard and soft sciences. Of course, as I have previously argued, one might wonder how a theory can renounce being a theory, and how observational instruments such as digital methods cannot still be methods, mediations, and representations.

In this context, I would prefer to 'import' such critique in the very structure of the digital. Latour believes that the digital is a good representation of social reality as he understands it because he sees the digital as a social network or a series of social networks. In his vision, he is certainly influenced by the structure of the Web 2.0. He is also influenced by the methods used by his collaborators at MédiaLab of Sciences Po Paris— for instance, web crawling and network visualization with tools such as Hyphe and Gephi. However, this no longer corresponds to the dominant aspects of the digital today. Sure, one might wonder if the digital has ever been flat, or if this was just the result of an illusion due to the interfaces that mask the existence of other levels or layers. But it is rather true that in the past what lay 'behind' the screen was economically, but also socially and culturally irrelevant. Everything changed when enterprises started collecting, exchanging (i.e. buy and sell), and efficiently analyzing data about users and consumers. We have assisted in what might be called a general 'algorithmization' and 'Big Datafication' of the digital.

My hypothesis is that in the last few years an algorithmic and Big Data superstructure has emerged, which plays an affirmative and even central role in our digital economy, culture, and society. It is precisely such a superstructure that a Latourian perspective ignores and would instead be at the core of a Bourdieusian one. The digital has become a Spaceland (in Abbot's novel, this is the name of the land where A Sphere lives), a tridimensional reality, in which the two-dimensional networked level of the social actants must be integrated to that of algorithms and Big Data analysis.

* * *

I would now like to introduce some reflections on what I call 'digital *habitus*.' As opportunely noticed by Héran (1987, 395–396), there is an intimate relation between the notions of *habitus* and schema, which is first of all etymological. The word "schema" used to indicate in the distant past the way of being, the bearing, the dressing or the costume— "schematotheque" was the accessory warehouse of a theater. In late antiquity, the term was used to indicate a monk's habit. Moreover, the words schema and *hexis* (which is the Greek term for *habitus*) come both from *echein*, which in Greek means "to have." From a morphological point of view, schema is to *echein* what *habitus* is to *habere*: an acquired result that the subject brings on and in herself, and hence a stable way of appearing to (and interacting with) others.[5] For this reason, I can say that my considerations on *habitus* are the continuation with other means (sociological and not philosophical anymore) of the reflections on the schematism I have been previously conducting. In both cases, it is a matter of the encounter and difficult articulation between passivity and activity, between what we receive, and what we give, between exteriority and interiority, generality and individuality, and of course between determinism and freedom.

The *habitus* is for Bourdieu a principle that acts at the level of the single subjects, of his or her actions and interactions with the world. However, the *habitus* is a supra- or infra-individual entity that frames our intentions without us even being conscious of such hetero-determination. The *habitus*, indeed, depends on the social group or class a subject belongs to. The *habitus* is what makes a social group or class become a group or a class; that is, what makes the single decisions and actions of each member of a social group or class, when it comes to specific objects and situations, resemble each other. In the words of the French sociologist, the *habitus* is a "conductorless orchestration which gives regularity, unity, and systematicity to the practices of a group or class, and this even in the absence of any spontaneous or externally imposed organization of individual projects" (Bourdieu 1977, 80). And again, "the practices of the members of the same group or class are more and better harmonized than the agents know or wish" (81).

It is noteworthy that for Bourdieu the *habitus* does not forge only actions and reactions, but also someone's desires and supposedly most authentic aspirations. To take a trivial example, one could say that the *habitus* frames the will of a young inhabitant of the Parisian banlieues of becoming a soccer player.[6] And it is for another *habitus* that a teenager of the same age living in one of the rich *rive gauche* districts of the French capital will rather desperately want to enter one of the *grandes écoles*. In summary, the *habitus* is what makes us want what society allows us to have. Once again, we are before a perversion of a Stoic principle, the one according to which humans should free themselves from certain passions that bring them to desire what they cannot reach. It is also noteworthy that the *habitus* determines for the French sociologist the kind of relations we have with each other: " 'Interpersonal' relationships are never, except in appearance, individual-to-individual relationships"; "the truth of the interaction is never entirely contained in the interaction" (81). To use another trivial example, one might say that it is not for individual characteristics, neither for a conscious attitude, but rather for a class or social group's *habitus* that the policeman beats the young adult of the Parisian banlieues while the young adult runs away from the police. In the Bourdieusian perspective, social actors do not even exist qua actors. They are rather there, and they interact with each other, as representatives of their respective class *habitus*. This perspective entertains some analogies with Dawkins's insights on the selfish gene. Finally, it is important to stress that the *habitus* is not just cognitively embedded, but also embodied in gestures, postures, movements, accents, and so on. Its transmission depends on fundamental institutions such as family and school. Actually—I will return to this point below—education has, for Bourdieu, the main function of sanctioning the reproduction of the social schemes, separations, and discriminations from generation to generation.

One of the most powerful descriptions of the *habitus* can be found in Bourdieu's critique of Sartre's voluntarism, for instance when he criticizes Sartre's description of a café waiter as someone who is playacting as a waiter, and for this reason is not choosing freedom but is rather consciously deceiving himself. According to Bourdieu, "the café waiter is not playing at being a waiter, as Sartre would have it. His body, which contains a history, *espouses* his job, in other words a history, a tradition, which he has never seen except incarnated in bodies, or more precisely, in the uniforms inhabited by a certain *habitus* that are called waiters" (Bourdieu 2000, 153–154). The Sartrian waiter, who freely chose his role and can or should freely choose to take distance from it, is for Bourdieu a social chimera, "a monster with a waiter's body and a philosopher's head." Incidentally, this does not mean that the philosopher is actually free, because the idea of freedom and total autonomy before a situation reflects in turn a very specific *habitus* to which Bourdieu refers to as the "scholastic illusion of distance from all positions" (154).

Now, there is already vast literature on Bourdieu, technology, and the digital. Stern (2003, 373) argues that for Bourdieu technology is not simply a thing filling a predetermined purpose. Rather, "technologies are socially shaped along with their meanings, functions, and domains of use." Technologies are for him little crystallized parts of *habitus*: "At a basic level, a technology is a repeatable social, cultural and material process [. . .] crystallized into a mechanism or set of related mechanisms" (376). For this reason, he sees continuity between Bourdieu and the Latourian notion of "inscription." However, from an authentic Bordieusian perspective, it is certainly more interesting to see how the social *habitus* impacts the relationships with and the uses of technologies. This is precisely the direction taken by most of the literature on Bourdieu and the digital—for a recent exhaustive overview, see Ignatow and Robinson (2017).

Several publications in this context concern, for instance, digital inequalities, the "informational" or "digital capital."[7] For instance, Jan van Dijk (in Ignatow and Robinson 2017, 952) defines the informational capital as the financial resources to pay for computers, but also the technical skills, the evaluation abilities, the information-seeking motivations, and so on, that essentially influence a subject's relationship with digital technologies. Other publications specifically resort to the "information *habitus*" or "digital *habitus*" for capturing how ways of interacting with digital technologies become habitualized by individuals operating within local social contexts and field positions. For example, Robinson (in Ignatow and Robinson 2017, 954) finds two types of information *habitus* in her ethnographic work on ICTs' use among low- and middle-income families in an agricultural belt in California. She distinguishes between a "taste of the necessary" among the low-income families and a "serious play" attitude among the upper-middle-income families. For the former, ICTs are considered in terms of their immediate practical value only. In the latter, children's use of the ICTs is encouraged because it is seen as promoting a deep level of technological engagement that will pay off in skills development over the long term. In all these cases, attention is oriented toward the consequences that social dynamics have on the use and the understanding of the digital. The digital is treated as an almost transparent means that reveals the consequences of non-technological conditions. My approach is different, insofar as I propose to understand the digital as a sort of *habitus* generator.[8]

In the present age of algorithms and Big Data, subjects are reduced to mere agglomerations of preferences, tendencies, and expected behaviors before a specific object, product, or situation. Consider, for instance, clustering, an unsupervised technique based on artificial intelligence algorithms. Dawn E. Holmes (2017, 22) offers a simple example based on credit card companies' practices:

> Clusters of profiles with similar properties are [. . .] identified electronically using an *iterative* [. . .] computer program. For example, a cluster may be defined on accounts with a typical spending range

or location, a customer's upper spending limit, or on the kind of items purchased, each resulting in a separate cluster. [. . .] When a new transaction is made, the cluster identification is computed for that transaction and if it is different from the existing cluster identification for that customer, it is treated as suspicious. Even if it falls within the usual cluster, if it is sufficiently far from the center of the cluster it may still arouse suspicion.

In this case, clustering has the function of finding an exception to the general rule. But of course, in most cases, the contrary happens: clustering and other techniques are used in order to find and confirm the rule. Interestingly enough, this methodology resembles some of the methodologies Bourdieu has privileged in the course of his career; such as the multiple correspondence analysis (MCA)—of Bourdieu's quantitative methods over the years; see Lebaron (2009). In both cases, one deals with the attempt of identifying patterns and groups of subjects with similar habits in terms of the consumption of goods, cultural tastes, and so on. There is for sure a sort of elective affinity between Bordieu's quantitative methods and his own theories about field, capital, and *habitus*. And one might suppose that there is also a family resemblance between Bourdieu's ideas and some of the dominant algorithmic and Big Data analysis practices today. My hypothesis, which should certainly be empirically tested, is that the reiterated contact of such a typified representation of oneself through online publicity, suggestions, search results, and so on, ends up in an embedment and embodiment of the digital *habitus*.

To sum up, one might say that two elements characterize the digital *habitus*: (1) first, the fact that the algorithmic and Big Data practices are indifferent to the subjects. They operate at an infra-individual level, insofar as they dismember individuals into some of their tendencies, tastes, et cetera. But they also operate at a super-individual level, because they reassemble these elements into general categories; (2) second, although indifferent to us, the digital is also extremely effective on us. It continuously proposes images and imaginaries of our identities and desires to which we unconsciously adapt to a certain extent, and that we make our own. However, there is also a fundamental difference between the Bourdieusian *habitus* and the digital *habitus*. Although often accused of determinism, Bourdieu's ultimate intention has never been deterministic. Rather, he used to believe that it is better to present freedom or autonomy as a result of a difficult path through hetero-determination than as an immediate act of courage or authenticity. For instance, in a passage of the *Pascalian Meditations* where he is criticizing Habermas's universalism and neo-enlightenment, Bourdieu (2000, 71) says that "[t] here is, appearance notwithstanding, no contradiction in fighting *at the same time against* the mystificatory hypocrisy of abstract universalism *and for* universal access to the conditions of access to the universal, the primordial objective which both universalistic preaching and nihilistic (pseudo-) subversion forget." For him, there is a radical difference between the desire of freedom and universality and its concrete realization. He also refers to a "thought about the social conditions of thought which offers thought the possibility of a genuine *freedom* with respect to those conditions" (118). Different is the case of the digital *habitus*. Indeed, with the tendency of the digital to flatten human subjects on their habitual behaviors and actions, to gather them in social groups and classes to better quantify and analyze them, there is, of course, no will of liberation but rather the contrary.[9]

* * *

I do not want to negate that the digital *habitus* can also be for good. I do believe that there are several good aspects in algorithms and Big Data analysis. Consider, for instance, the possibility of lowering diagnostic errors by 85 percent in breast cancer patients (Taddeo and Floridi 2018, 751).[10] Still the fact remains that the digital has its risks and perversions that must be opportunely analyzed and criticized.

The question I want to ask at this point is how it is possible to make subjects able to deal with the digital *habitus* in order to carve out room for maneuvering or allowing a margin of freedom before the power and the configuring force exercised on them by and through the sociotechnical systems. I have the impression that two types of answers can be given. The first one is subjectivistic and even existentialistic in some respects, while the second one is rather social and institutional. My hypothesis is that the latter must be considered as a condition of the possibility of the former.

In order to explain the subjectivistic vision, I am going to refer to Foucault. Two Foucauldian notions have played an important role in the field of critical digital studies. In Romele et al. (2017) we have insisted on Panopticism, which over the years found breeding grounds in Internet and social media studies. Academics used terms such as "dataveillance," "superpanopticon," "panoptic sort," and "electronic panopticon." Researchers used to consider social media as technologies of mutual control and surveillance. A sort of reversed Panopticon in which the controlled is alone in the middle of the prison and the controllers are all around. The more recent centralization of the Web around few big enterprises and the awareness of mass surveillance programs such as PRISM brought researchers to consider the Internet and social media as more classic forms of Panopticon. For them, the problem is not the relationships among individuals *in* a sociotechnical system, but rather the relations between individuals *and* the sociotechnical system.

Confession is another Foucauldian notion that has exercised fascination among digital media and technologies scholars. Incidentally, Foucault did not use the French term *confession*, but the more neutral *aveu*.

The former has a stronger religious connotation, while the latter is closer to a judicial one. In the first volume of *The History of Sexuality*, Foucault (1978, 21) says that "[t]he Christian pastoral prescribed as a fundamental duty the task of passing everything having to do with sex through the endless mill of speech." He affirms that "the Christian pastoral also sought to produce specific [. . .] effects of displacement, intensification, reorientation and modification of desire itself" (23). With the rise of Protestantism, the Counter Reformation, eighteenth-century pedagogy, and nineteenth-century medicine, confession has spread far beyond the limits of Catholicism, and today, one can say that "Western man is a confessing animal" (59)—"*bête d'aveu*" in the French expression. Foucault is criticizing the idea, according to which, speaking of one's own desires has liberating effects per se; as it was commonly believed during and after the sexual revolution in 1960s. In particular, he wants to show how, contrary to the opinion that the discourse about sex and sexuality had been removed from society in the past, the Victorian society had displaced a fine array of practices for making people talk about sex and sexuality in highly codified contexts such as schools, hospitals, and so on. The French philosopher is not criticizing the possibility and opportunity of speaking one's own mind about sexuality and other intimate matters, but he seems more concerned about the places and contexts in which such 'confessions' take place. One can suppose that for him all externalizations of this kind generate power relations, but power is not necessarily bad. It can also be intriguing, and the problems arise when power becomes a source of domination. In a sentence that recalls the figures of a psychoanalyst, Catholic priest, and judge, he says that within confession "the agency of domination does not reside in the one who speaks [. . .] but in the one who listens and says nothing; not in the one who knows and answers, but in the one who questions and is not supposed to know" (72).

Digital media and technologies might be considered as further developments of the Western confessional machine beyond the limits of Western societies and institutionalized powers. For instance, it can be argued that Facebook's reaction buttons induce users to 'confess' their sentiments, while intensifying and disciplining (and hence monetizing) them. For this reason, Boellstorff (2013) states, from a Foucauldian perspective, confession is the master metaphor for explaining Big Data: "confession is a modern mode of making data, an incitement to discourse we might now term *an incitement to disclose*. It is profoundly dialogical: one confesses to a powerful Other." For Friesen (2017), one can consider social media such as Facebook "as a powerful, interpellating, confessional technology of the self." In particular, he distinguishes among two levels of interpellation in digital media. The first one is characterized by familiarity, simplicity, and the use of the dialogic form, when Facebook, for instance, greets its users with the question: "what's on your mind?" The second is instead characterized by the categorization of its users "according to standardized, predefined classifications." These two levels recall the three-dimensionality of the digital I have previously discussed.

There is, of course, no reason to disagree with these two ideas. The digital today is determined by both new forms of governmentality and incitement toward disclosure. However, as I suggested in the conclusion of Romele (2018), Panopticism and confession may be integrated by a reflection on a third, again Foucauldian, paradigm: *parrhesia*. While early Foucault analyzed the ways in which the modern subject has been shaped by technologies of power, in his later work he attempted to develop an 'ethics' or 'aesthetics' of the self "as the active engagement of people with governing and fashioning their own way of being in relation to conditioning circumstances" (Dorrestijn 2012, 227). In other words, while the earlier Foucault is concerned about *interpreting* modern subjectivations, the later is more interested in finding ways of *changing* or at least dealing with them.

Foucault (2011a, 55) defines *parrhesia*, an ancient Greek term literally meaning "to speak everything" and by extension, "to speak freely" or "candidly," as a way of telling the truth that "does not fall within the province of eristic and an art of debate, or of pedagogy and an art of teaching, or of rhetoric and an art of persuasion, or of an art of demonstration." There is a crucial difference, for him, between the *parrhesia* and a performative utterance. In the latter, the effect, which is known in advance, is codified. *Parrhesia*, instead, does not produce a codified effect, but it opens up an unspecified risk. Although Foucault never refers to them, narrative and productive imagination can be seen as forms of *parrhesia* since they have to do with the Socratic *"didonai logon,"* an expression that Foucault himself (2011b, 159) opportunely renders as "to give an account of themselves, to explain themselves."

Now, giving narratively an account of oneself before others is a possibility that in the age of Big Data and new algorithmic machines is not denied, but anesthetized or made inoperative. There is no need to confess because any powerful sociotechnical system is capable of predicting one's choices and actions on the basis of her present and past behaviors. The relevance of the stories someone tells about herself will always fail before the exhaustivity of the information a digital machine has at its disposal, and its capacity to correlate. "Who needs stories if you can get the data?" asks Mireille Hildebrandt:

> We may assume that statistical inference could one day refuse the stories a person tells to give an account of herself. At some point the police, the judge, the insurance agent, your webshop, Google Analytics, your doctor and your tax collector will point out that you match an aggregated profile that probably (sic!) provides a better prediction of your future behavior than whatever you tell them.
>
> (Hildebrandt 2011, 388)

Notice that this does not mean that Big Data and algorithmic machines actually *make* better predictions about oneself. One may say that at the end of the day it does not really matter if digital technologies are capable of good predictions about our behaviors—although, of course, they increasingly are. What is even more important is that people, and especially those who make decisions in fields such as health and security, *do* increasingly delegate these decisions to the digital machines, because they *do* trust their emerging capacities.

Thus, it can be said that any form of speaking candidly, and taking risks in doing so, is increasingly anesthetized. If one assumes that *parrhesia* is a truth-telling that creates rupture and novelty, a force representing the possibility, capacity, and capability to negotiate with a specific power relation, it is precisely this potentiality that becomes irrelevant. Before such a force of prediction, and the human trust in this force, no individual reaction, however exemplary, can be effective. This is, for me, the main limitation of the later Foucault's ethics and aesthetics of the self, once applied to the digital. And this might be more widely considered as the limit of all individualistic or communitarian approaches to the issue. I am referring to all forms of 'exit' (e.g. unsubscription, technology-free vacations, et cetera), 'counterattack' (e.g. tracking the trackers), or 'deviating' (e.g. hacking) strategies concerning the digital, which can lead to forms of elitism and paternalism at best. For the same reason, I have a mixed opinion about Shannon Vallor's notion of "technomoral virtue ethics." On the one hand, she opportunely argues for instance that "satisfactory solutions to digital media 'stickiness' will have to involve *collective* cultural agreements to seek healthier digital norms and social rituals" (Vallor 2016, 169). On the other hand, she remains within an ethic tradition which is in great part individualistic and elitist, and tends to neglect the sociopolitical roots of digital inequalities. In some sense, I would say that technosociopolitics must be considered as the condition of possibility for a technomoral virtue ethics à la Vallor.

According to Sloterdijk (2013, 181), the greatest weakness of the *habitus* concept is that "it cannot grasp the individualized forms of existential self-designs. Bourdieu's analysis necessarily remains within the typical, the pre-personal and the average, as if the Homo sociologicus were to have the last word on all matters." He consequently undertakes a reinterpretation of the philosophical notion of *habitus* stressing that in ancient and medieval tradition, in Aristotle and Aquinas, the *habitus* (*hexis* in Greek) had to do with the "virtuous within us," and even with the "good within us." It seems to me that Sloterdijk does not really understand Bourdieu's real perspective, which consists in making room for the possibility of a *habitus* à la Aristotle or Aquinas—although certainly less personalistic, heroic, and ultimately aristocratic than what Aristotle, Aquinas, but also Heidegger, Sartre, and Sloterdijk himself has thought as philosophers. Chantal Jaquet (2014, 219) opposes the Bourdieusian *habitus* to what she calls "complexion." She affirms that the "transclasses" (people that moved from one social class of origin to another) are not heroes, and that the phenomenon of "non-reproduction" is not individual, but transindividual. This means that the 'success' of the transclass is not the result of a radical rupture with their origin and with the power relations that have constituted them as a self, but rather the result of a reconfiguration, a new arrangement of them. The term "complexion" refers precisely to the chain of hetero-determinations that articulate each other in order to create the weft of a single life. Such notion helps in rejecting both an essentialist (that is, deterministic) and an existentialistic perspective on human beings. The non-reproduction is less a self-production than a social co-production in which one learns to deal with her social background and present milieu. The transclass is the one who clears her own passage between multiple social forces. This somehow recalls Peter-Paul Verbeek's (2011, 85) postphenomenologic interpretation of the later Foucault, according to which individuals can positively transform their condition, although remaining within a power relation. However, instead of opposing complexion and *habitus*, I propose to consider Bourdieu's reflections and analysis of the social *habitus* as the condition of possibility of a complexion which is more than just the result of a series of fortuitous events and encounters in someone's life.

However, the question arises as to how it is possible not only to observe and eventually criticize the social *habitus* but also to change it. In the first part of her book, Jaquet (2014, 23–102) proposes a list of the causes of non-reproduction: ambition, models and mimetism (familiar model, scholastic model, et cetera), socioeconomic and institutional initiatives, affects and encounters, and the role of the social milieu that for instance rejects some of its members like homosexuals. Yet, among these, only the socioeconomic and institutional initiatives seem to me to have the potential to escape the individual and occasional chance to which the task of developing one's own moral virtues is too often entrusted. As I have previously stated, education is for Bourdieu the most powerful means of reproduction of the social schemes, separations, and discriminations from generation to generation. However, Jean-Claude Passeron (in Emmenegger 2016, 65), the co-author with Bourdieu of *The Inheritors*, declared that their intention in that book was not to negate the emancipatory potential coming from democratization of education. Rather, they wanted to suggest the possibility of constructing a "rational pedagogy" aiming at dismantling the mechanism of sociocultural reproduction through specific attention to social inequalities. If education is the most effective means of reproduction, could not we think that it could also play an essential role in the non-reproduction?

This perspective can, in my opinion, be easily imported into the debate about the digital. I am thinking, of course, about exhorting and teaching people how to protect themselves, and how to achieve, through the digital means, their own expectations and those of their communities. But I am also thinking, more widely, to the intervention of national and international institutions in mediating and harmonizing the relationships between individuals or communities and the sociotechnical systems, with the precise scope of empowering these individuals or communities.

In critical theory, institution usually rhymes with coercion and illusion. But the recent history of the digital in Europe is, for instance, also characterized by the capacity and the will of some institutions to embrace individual and exemplary initiatives. Let us think about the legal initiatives of Costeja González and Max Schrems that respectively brought us to the Right to Be Forgotten Ruling, and to the declaration by the Court of Justice of the European Union of invalidity for the Safe Harbor framework. Of course, I am not saying that these are the best examples possible. Nor I am saying that institutions are per se enough. Indeed, institutions can also explicitly refuse to embrace such initiatives. Furthermore, institutions can work to the detriment of the ones they are supposed to represent, defend, and protect. I would say that institutions also need to be just. I am referring to Ricoeur's (1992, 172) ethical intention as "aiming at the 'good life' with and for others, in just institutions." I am aware that the idea of 'just institutions,' in a digital milieu as well as in general, would deserve further discussion. Here, under the guise of a conclusion, I limit myself to say that the goal of such a form of justice would consist of creating sociotechnical conditions for an *ethos* of distanciation from one's own digital *habitus*. In other words, it would mean to contribute to framing a sociodigital environment in which people can become sensitive to the insensibility and indifference of the digital.

Notes

1. This "criteriological" meaning is just one of the meanings Kearney attributes to the term, the others being "critical," "grammatological," "diagnostic," and "carnal."
2. I am particularly grateful to Camilla Emmenegger who with her M.A. dissertation "La critica come distinzione: Il pensiero di Pierre Bourdieu" (The critics as distinction: The thought of Pierre Bourdieu) discussed in 2016 at the University of Turin (Italy) made me discover the philosophical relevance of Bourdieu's sociology.
3. B. Latour, "Pour être de gauche, il ne suffit pas de parler des dominés. Ni de donner des leçons de morale aux pouvoirs au nom d'une sociologie. La gauche a-t-elle besoin de Bourdieu?" www.liberation.fr/tribune/1998/09/15/pour-etre-de-gauche-il-ne-suffit-pas-de-parler-des-domines-ni-de-donner-des-lecons-de-morale-aux-pou_245845. *Libération*, September 15, 1998. Accessed June 10, 2019.
4. The difference and the rivalry between Latour's and Bourdieu's social ontologies and methodologies recalls the one between Tarde's and Durkheim's. As it emerges from a reconstruction of a debate between the two sociologists that took place in 1903 at the École des Hautes Études Sociales in Paris, the divergence was mainly on two points. First, on the epistemological question

if and how it was possible to constitute sociology as a science having clear methodologies, different then from philosophy. For Durkheim, if sociology "wishes to live up to the hopes which have been built around it, it must strive to become more than an eccentric kind of philosophical literature." But for Tarde it is absurd to think of an absolute autonomy of sociology from other forms of knowledge: "Not everything that members of a society do is sociological. [. . .] To breathe, digest, blink one's eyes, move one's leg automatically, look absently at the scenery or cry out inadvertently, there is nothing social about these acts." Second, the ontological issue consisting into knowing if it is individuals and their multiple relations that make society or there are rather general laws and regularities that cannot be reduced to the individuals. According to Durkheim, there is a common sense for which "social life can have no other substratum than individual consciousness; otherwise it appears to be up in the air, floating in an empty space." But things are actually different: as life emerges from inanimate matter but cannot be reduced to it, as consciousness emerges from the interactions between neurons but is something substantially different from it, so the social reality is an emergence of the individual actions and interactions, but it is also something essentially more than them. Tarde accuses instead his interlocutor of inventing forces that have an existence "independent of human personality, and rule man with despotic might, by the oppressive shadow which they cast over him." It is noteworthy, although not surprising, that in the 2007 performed and recorded reconstruction of the debate in French, Latour plays the part of Tarde (while Durkheim is played by Bruno Karsenti). See www.bruno-latour.fr/fr/node/424. Accessed June 10, 2019. A transcribed version in English is available at www.bruno-latour.fr/sites/default/files/downloads/TARDE-DURKHEIM-GB.pdf. Accessed June 10, 2019.

5. According to this author (Héran 1987, 398–399) the verb "to have" has the specificity of being neither active nor passive. Despite its transitive structure ("Pierre has a house"), it is just a pseudo-transitive verb because it cannot be used passively ("A house is had by Pierre"). The fact is that this verb at the origin did not use to indicate a process between an acting subject and something external to her, but a specific relation that the subject entertains with her own states, dispositions, and attitudes—or with some intimate elements such as clothing, wounds, and sicknesses. The possession of an external object is just a derivative meaning. He also observes that the construction of past tenses with the verb "to have" as auxiliary used to indicate precisely the possession of an individual *habitus* constituted in the course of past experiences of perception and judgment. Just later it has been applied to all other kinds of experiences.

6. Arsene Wenger, former manager of Arsenal, famously ranked the Greater Paris region as the second-best talent pool in soccer after Sao Paolo. Many argue that the Paris suburbs rank now top. See, for instance, S. Kuper, "From Pogba to Mbappe: Why Greater Paris Is the World's Top Talent Pool." www.espn.com/soccer/blog/espn-fc-united/68/post/3320634/greater-paris-biggest-talent-pool-in-football-paul-pogba-kylian-mbappe-anthony-martial. *ESPN.com*, December 27, 2017. Accessed June 10, 2019.

7. "Capital" and "field" are the two other fundamental notions of Bourdieu's sociology. Bourdieu uses the term "field" in order to describe groups of interrelated actors, and "capital" to describe the specific forms of agency and prestige within a specific field.

8. For a similar perspective, see Fourcade and Healy (2016, 8), who introduced the notion of "übercapital," that is, "a form of capital arising from one's position and trajectory according to various scoring, grading, and ranking

methods"—many of them of course related to the ubiquitous presence of connected digital devices.

9. One could say that the digital *habitus* is more fragmented and localized, and hence less totalizing than the Bourdieusian *habitus*. Moreover, one might argue that the fact that the digital *habitus* is materialized makes it in some sense visible, and hence more easily criticizable. Just to take an immediate example, consider the possibility offered by Facebook of exploring and changing one's own ad preferences—"your interests," "advertisers you've interacted with," "ad settings," et cetera. However, I believe that these same elements (fragmentation and materialization) might actually increase the effectiveness of this *habitus*.

10. In the words of these authors, "[w]ith their predictive capabilities and relentless nudging, ubiquitous but imperceptible, AI systems can shape our choices and actions easily and quietly. This is not necessarily detrimental. For example, it may foster social interaction and cooperation" (Taddeo and Floridi 2018, 751).

References

Boellstorff, T. 2013. "Making Big Data: In Theory." *First Monday* 18(10). http://firstmonday.org/article/view/4869/3750. Accessed June 10, 2019.

Bourdieu, P. 2000. *Pascalian Meditations*. Stanford: Stanford University Press.

———. 1977. *Outline of a Theory of Practice*. Cambridge: Cambridge University Press.

Braidotti, R. 2016. "Posthuman Critical Theory." In D. Banerji, and M.R. Paranjape (eds.). *Critical Posthumanism and Planetary Futures*. Dordrecht: Springer, 13–32.

Dorrestijn, S. 2012. "Technical Mediation and Subjectivation: Tracing and Extending Foucault's Philosophy of Technology." *Philosophy & Technology* 25(2): 221–241.

Emmenegger, C. 2016. *La critica come distinzione: Il pensiero di Pierre Bourdieu*. M.A. Dissertation, University of Turin (unpublished).

Foucault, M. 2011a. *The Courage of Truth: Lectures at the Collège de France (1983–1984)*. London: Palgrave Macmillan.

———. 2011b. *The Government of Self and Others: Lectures at the Collège de France (1982–1983)*. London: Palgrave Macmillan.

———. 1978. *The History of Sexuality: Volume 1: An Introduction*. New York: Pantheon Books.

Fourcade, M., and Healy, K. 2016. "Seeing Like a Market." *Socio-Economic Review* 15(1): 9–29.

Friesen, N. 2017. "Confessional Technologies of the Self: From Seneca to Social Media." *First Monday* 22(6). http://firstmonday.org/ojs/index.php/fm/article/view/6750/6300. Accessed June 10, 2019.

Héran, F. 1987. "La seconde nature de l'habitus. Tradition philosophique et sens commun dans le langage sociologique." *Revue française de sociologie* 28(3): 385–416.

Hildebrandt, M. 2011. "Who Needs Stories If You Can Get the Data?" *Philosophy & Technology* 24(4): 371–390.

Holmes, D.E. 2017. *Big Data: A Very Short Introduction*. Oxford: Oxford University Press.

Ignatow, G., and Robinson, L. 2017. "Pierre Bourdieu: Theorizing the Digital." *Information, Communication & Society* 43(2): 197–210.

Jaquet, C. 2014. *Les transclasses ou la non-reproduction.* Paris: P.U.F.

Kearney, R. 2011. "What Is Diacritical Hermeneutics?" *Journal of Applied Hermeneutics* 10: 1–14.

Latour, B. 2005. *Reassembling the Social: An Introduction to Actor-Network Theory.* Oxford: Oxford University Press.

Lebaron, F. 2009. "How Bourdieu 'Quantified' Bourdieu: The Geometric Modelling of Data." In K. Robson, and C. Sanders (eds.). *Quantifying Theory: Pierre Bourdieu.* Dordrecht: Springer, 11–29.

Pinto, Louis. 2009. *Le café du commerce des penseurs: A propos de la* doxa *intellectuelle.* Paris: Éditions du Croquant.

Ricoeur, P. 1992. *Oneself as Another.* Chicago: The University of Chicago Press.

Romele, A. 2018. "Herméneutique (du) digital: les limites techniques de l'interprétation." *Études Digitales* 3: 55–74.

Romele, A., Gallino, F. Emmenegger, C., and Gorgone, D. 2017. "Panopticism Is Not Enough: Social Media as Technologies of Voluntary Servitude." *Surveillance & Society* 15(2): 204–221.

Sloterdijk, P. 2013. *You Must Change Your Life.* Cambridge: Polity Press.

Stern, J. 2003. "Bourdieu, Technique and Technology." *Cultural Studies* 17(3–4): 367–389.

Taddeo, M., and Floridi, L. 2008. "How AI Can Be a Force for Good." *Science* 361(6404): 751–752.

Vallor, S. 2016. *Technology and the Virtues: A Philosophical Guide to a Future Worth Wanting.* Oxford: Oxford University Press.

Verbeek, P.P. 2011. *Moralizing Technology: Understanding and Designing the Morality of Things.* Chicago: The University of Chicago Press.

Index

Printed in the United States
by Baker & Taylor Publisher Services

Printed in the United States
by Baker & Taylor Publisher Services